A METAPHYSICS FOR SCIENTIFIC REALISM

Scientific realism is the view that our best scientific theories give approximately true descriptions of both observable and unobservable aspects of a mind-independent world. Debates between realists and their critics are at the very heart of the philosophy of science. Anjan Chakravartty traces the contemporary evolution of realism by examining the most promising recent strategies adopted by its proponents in response to the forceful challenges of antirealist sceptics, resulting in a positive proposal for scientific realism today. He examines the core principles of the realist position, and sheds light on topics including the varieties of metaphysical commitment required, and the nature of the conflict between realism and its empiricist rivals. By illuminating the connections between realist interpretations of scientific knowledge and the metaphysical foundations supporting them, his book offers a compelling vision of how realism can provide an internally consistent and coherent account of scientific knowledge.

ANJAN CHAKRAVARTTY is Associate Professor at the Institute for the History and Philosophy of Science and Technology and Department of Philosophy, University of Toronto.

A METAPHYSICS FOR SCIENTIFIC REALISM

Knowing the Unobservable

ANJAN CHAKRAVARTTY

University of Toronto

CAMBRIDGE
UNIVERSITY PRESS

CAMBRIDGE UNIVERSITY PRESS
Cambridge, New York, Melbourne, Madrid, Cape Town, Singapore,
São Paulo, Delhi, Dubai, Tokyo

Cambridge University Press
The Edinburgh Building, Cambridge CB2 8RU, UK

Published in the United States of America by Cambridge University Press, New York

www.cambridge.org
Information on this title: www.cambridge.org/9780521876490

First published 2007

A catalogue record for this publication is available from the British Library

ISBN 978-0-521-87649-0 Hardback
ISBN 978-0-521-13009-7 Paperback

Transferred to digital printing 2009

It is the fault of our science that it wants to explain all; and if it explain not, then it says there is nothing to explain.

Van Helsing to Dr Seward
Bram Stoker, *Dracula*

Contents

Tables

ix

Figures

Preface

This begins as a book about scientific realism. To a very rough, first approximation, realism is the view that our best scientific theories correctly describe both observable and unobservable parts of the world. When philosophers consider this idea they are usually concerned to address the issue of whether it is a reasonable view to hold. They worry about whether it gives a plausible account of scientific knowledge, and rightly so! This is an undeniably important question. It is close to the heart of almost all issues in the philosophy of science, and importantly relevant to many issues in philosophy and the sciences more generally. This book, however, starts with a much more basic and arguably prior question. What is scientific realism, exactly?

One might think that in order to discuss the question of whether realism is plausible or reasonable, one should already know what it is. As philosophers know only too well, however, one cannot think about everything at once, and the debate surrounding realism is no exception. One must often assume coherent accounts of various components of a position in order to give careful attention to others, and people on all sides of this debate usually take a great deal for granted so as to focus on epistemic questions. For example, when describing their positions realists often rely heavily on things such as causation, laws of nature, and the natural kind structure of the world. These ontological ingredients play important roles in disputes about realism, but the natures of these things are generally passed over *tout court* in these disputes specifically. Their brief mention leaves open the question of whether such metaphysical foundations are themselves secure enough or otherwise appropriate to support the edifice of realism.

So, what begins as a book about scientific realism soon becomes a book about its foundations, and as a consequence this work is not a defence of realism, *per se*. Nevertheless, equipped with a better understanding of what a view entails and does not entail, one may find oneself in a better position to defend or condemn it. I believe this to be the case here, and though my primary objective is not to defend or to condemn, I hope that a clarification of what realism entails will facilitate further discussion of the important disputes between realists and antirealists. Currently, much of what can be said regarding some of these disputes *has* been said, and in order to move forward perhaps greater clarity is needed regarding the nature of this world which realism takes to be illuminated by the sciences. The metaphysics of realism has lagged behind its epistemology, and one of the best reasons for addressing the former is to facilitate better the latter. But the metaphysics of realism comprises a fascinating set of issues on its own, and in this book I aim to consider them.

Some think there are as many versions of scientific realism as there are scientific realists. That is probably a conservative estimate! There are probably as many versions of realism as there are realists and antirealists. What hope is there, then, for a book about what scientific realism *is*, let alone a proposal for a metaphysics supporting it? It would certainly be impossible to describe realism precisely in a way that would satisfy all realists and antirealists. No one detailed account answers the descriptive question of what scientific realism is. That said, I believe there is something like an account (with negotiable boundaries) that answers the descriptive question of what some of the best hopes for conceiving realism may be. Certain elements of realist views appear time and again in divergent accounts, and their recurrence suggests their centrality to realist approaches generally. These commonalities merit attention on any version of realism. No discussion of these matters can hope to be purely descriptive, of course, and arguments for the many normative suggestions I will make concerning what I take to be the most promising ways to understand realism appear throughout. Many will surely disagree with the account of realism argued for here, but I do hope that realists and antirealists alike will take an interest in the arguments for why the realism I describe is attractive and defensible.

These arguments serve two distinct but closely related ends. The first is to identify metaphysical commitments that are importantly constitutive of realism, and thus crucial to the internal coherence of the position. Thinking about these commitments helps one to distinguish them from others that fall outside the immediate context of realism, and to clarify the ways in which they do and do not conflict with traditional empiricist rivals. Importantly, though – and I cannot stress this enough – I will not argue for the exclusive coherence of the metaphysical account I propose. For reasons discussed in Chapter 1, I suspect that different metaphysical approaches are consistent with realism and comprise a spectrum, from

Humean austerity regarding certain metaphysical questions to the more elaborate terrain I will map and beyond. The differences between these approaches have consequences for the sorts of things realists may hope to explain, and I endeavour to clarify various trade-offs involving ontology and explanation throughout. The second objective of this work is to give a unified account of a metaphysical proposal in support of realism, and here in particular various normative suggestions take centre stage. To summarize the aims of the book very concisely, I investigate the core elements of promising versions of contemporary realism, and develop a metaphysics that makes sense of these commitments. The end product, I hope, is a basic framework with a capacity for elaboration by realists and antirealists both, as may be appropriate to the specific issues they engage.

Here is a brief description of the contents by chapter. The first part of the book, 'Scientific realism today', sketches a preliminary account of the central commitments of realism as they have evolved over time and quite recently, often in response to antirealist scepticism. The essence of the controversy between realists and antirealists concerns the possibility of having knowledge of the unobservable, and this possibility is most strongly contested by varieties of empiricism. In Chapter 1, I introduce the idea of realism in the context of the sciences and consider the dialectic between this position and the forms of empiricism that dispute it, thus illuminating some different senses in which realism is apparently metaphysical. In Chapter 2, I develop this initial sketch of realism by considering what I take to be its most promising formulations, such as entity realism and structural realism, in order to produce a portrait of the position that incorporates the best of their insights and avoids their defects. The resulting inventory of realist commitment, to certain properties, relations, and particulars, and various connections between them are explored in Chapter 3.

'Metaphysical foundations', the second part of the book, delves more deeply into the ontological issues raised by the contemporary view of realism offered in Part I. The internal coherence of realism depends in part on the possibility of articulating an integrated and compelling account of these issues, and I articulate one such account here. Chapter 4 examines the issue of causation, on which much of the justificatory story of realism depends, and argues that its role in this story is nicely facilitated by a specific understanding of causal phenomena in terms of processes and dispositions. This discussion is extended in Chapter 5 to a consideration of laws of nature, where I argue that the natures of causal properties and the dispositions they confer lend themselves to a promising and metaphysically minimal account of natural necessity. Moving from an investigation of properties and relations to objects, Chapter 6 focuses on the role played in realist discourse by the concept of natural kinds. I argue that a proper understanding of this concept results in a dissolution of the traditional dichotomy between objective and subjective classification, and a rejection of certain vestiges of ancient metaphysics, outmoded in the context of realism today.

In the final part of the book, 'Theory meets world', I consider several matters arising from Parts I and II that overlap the hazy boundary between the metaphysics and the epistemology of realism. In Chapter 7, I examine the use of models to represent parts of the world, and the question of whether the "ontological" nature of scientific theories, conceived either linguistically or in terms of models, has any bearing on the epistemic commitments of realism. Chapter 8 builds on this discussion by giving an amalgamated account of certain features of theories and models that have implications for a realist understanding of scientific knowledge. Drawing analogies to representation in art, these features include the use of abstraction and idealization, and the notion of approximate truth.

It is sometimes said that scientific realism is a perennial issue of philosophy. Indeed, one of the implicit themes of this book is that some disputes between realists and antirealists, not to mention disputes between realists with different philosophical predispositions, are destined to remain unresolved due to an irresolvable lack of shared assumptions. To a great extent, these assumptions concern the metaphysical aspects of realism. It seems unlikely to me that there are convincing responses to all forms of antirealist scepticism, and it seems even less likely that there are any knock-down arguments against them. Some forms of scepticism are, no doubt, coherent philosophical positions, and it is doubtful whether there are any non-question-begging arguments that will decide these matters ultimately. If one is interested in realism, however – in seeing whether *it* can be understood as an engaging, coherent, compelling account of the sciences – then much work remains to be done. If one feels any pull in this direction, then it is crucial that one have recourse to an internally consistent and substantive position. It is the goal of this book to furnish a unified picture of the metaphysics of scientific realism with which to answer this challenge. It aspires to give a wide-ranging answer to the question of what sort of realist a sophisticated realist can be.

Much of the book took shape while I was a Visiting Fellow at the Center for Philosophy of Science at the University of Pittsburgh in 2004. I am grateful for that opportunity, as well as to the Social Sciences and Humanities Research Council of Canada for financial support. I am also thankful to the following journals and publishers for permission to make use of work published previously. Sections 1.4 and 1.5 extend arguments found in 'Stance Relativism: Empiricism versus Metaphysics', *Studies in History and Philosophy of Science* 35: 173–84, © 2004 Elsevier B. V., all rights reserved. Sections 2.2–2.5 are based on material in 'Structuralism as a Form of Scientific Realism', *International Studies in the Philosophy of Science* 18: 151–71, © 2004 Taylor & Francis Group, all rights reserved. Section 3.2 is a reworking of parts of 'Semirealism', *Studies in History and Philosophy of Science* 29: 391–408, © 1998 Elsevier B. V., all rights reserved, and sections 3.3–3.5 are adapted from 'The Structuralist Conception of Objects', *Philosophy of Science* 70: 867–78, © 2003 Philosophy of Science Association, all rights reserved. Chapter 4 is based on 'Causal Realism: Events and Processes', *Erkenntnis*: 63: 7–31, © 2005 Springer Science & Business Media, all rights reserved, and Chapter 5 is adapted from 'The Dispositional Essentialist View of Properties and Laws', *International Journal of Philosophical Studies* 11: 393–413, © 2003 Taylor & Francis Group, all rights reserved. Sections 7.2–7.5 rework arguments found in 'The Semantic or Model-Theoretic View of Theories and Scientific Realism', *Synthese* 127: 325–45, © 2001 Springer Science & Business Media, all rights reserved.

Before getting under way there are also a number of people I must thank. I have benefited immensely from discussions with them, and all of the following were kind enough to read earlier drafts of some part of this material at one stage or another. I owe a great debt to David Armstrong, Alexander Bird, Simon Bostock, Bryson Brown, James Robert Brown, Otávio Bueno, Jeremy Butterfield, Krister Bykvist, Pierre Cruse, Thomas Dixon, Brian Ellis, Jason Grossman, Anandi Hattiangadi, Katherine Hawley, Eric Heatherington, Nick Jardine, Jeff Ketland, Martin Kusch, James Ladyman, Tim Lewens, Gordon McOuat, Hugh Mellor, Stephen Mumford, Robert Nola, Stathis Psillos, Michael Redhead, Michael Rich, Juha Saatsi, Howard Sankey, Peter Smith, Kyle Stanford, Paul Teller, Martin Thomson-Jones, and Bas van Fraassen. I am very grateful to Matthias Frisch and to a reader for Cambridge University Press, whose excellent comments on the entire manuscript and attention to detail led to many improvements, to Hilary Gaskin and Jo Breeze at the Press for steering this project through so helpfully, to Jo Bramwell for her copy-editing skills, and to Steve Russell for the index. The incredible generosity and thoughtfulness of Peter Lipton, Steven French, and Margie Morrison have been an unremitting inspiration. Thanks go to my family for their support: my parents, my sister, the Gangulys, the Jacksons, and the gang.

Three final remarks are in order concerning the chapters to follow. Single quotation marks indicate quotation, or the mention of a term or phrase. Double quotation marks indicate the (generally figurative or metaphorical) use of a term or phrase. Lastly, a warning: I have made extensive reference to many tempting desserts. The reader is advised to snack before reading.

Abbreviations

DIT   dispositional identity thesis, for causal properties
ER   entity realism
IBE   inference to the best explanation, sometimes called abduction
NE   the New Essentialism, concerning scientific ontology
NOA   the natural ontological attitude
PI   pessimistic induction, or pessimistic meta-induction
PII   principle of the identity of indiscernibles
QM   quantum mechanics
SD   Salmon-Dowe (causal process)
SR   structural realism
UTD   the underdetermination of theory choice by data or evidence

PART I

Scientific realism today

CHAPTER I

Realism and antirealism; metaphysics and empiricism

1.1 THE TROUBLE WITH COMMON SENSE

Hanging in my office is a framed photograph of an armillary sphere, which resides in the Whipple Museum of the History of Science in Cambridge, England. An armillary sphere is a celestial globe. It is made up of a spherical model of the planet Earth (the sort we all played with as children), but the model is surrounded by an intricate skeleton of graduated rings, representing the most important celestial circles. Armillary spheres were devised in ancient Greece and developed as instruments for teaching and astronomical calculation. During the same period, heavenly bodies were widely conceived as fixed to the surfaces of concentrically arranged crystalline spheres, which rotate around the Earth at their centre.

This particular armillary sphere has, I expect, many fascinating historical stories to tell, but there is a specific reason I framed the picture. Once upon a time, astronomers speculated about the causes and mechanisms of the motions of the planets and stars, and their ontology of crystalline spheres was a central feature of astronomical theory for hundreds of years. But crystalline spheres are not the sorts of things one can observe, at least not with the naked eye from the surface of the Earth. Even if it had turned out that they exist, it is doubtful one would have been able to devise an instrument to detect them before the days of satellites and space shuttles. Much of the energy of the sciences is consumed in the attempt to work out and describe things that are inaccessible to the unaided senses, whether in practice or in principle. My armillary sphere, with its glorious and complicated mess of interwoven circles, is a reminder of past testaments to that obsession.

In describing the notion of a crystalline sphere, I have already made some distinctions. There are things that one can, under favourable circumstances, perceive with one's unaided senses. Let us call them "observables", though this is to privilege vision over the other senses for the sake of terminological convenience. Unobservables, then, are things one cannot perceive with one's unaided senses, and this category divides into two subcategories. Some unobservables are nonetheless detectable through the use of instruments with which one hopes to "extend" one's senses, and others are simply undetectable. These distinctions are important, because major controversies about how to interpret the claims of the sciences revolve around them. In this chapter, I will briefly outline the most important positions engaged in these controversies, and consider how the tension between speculative metaphysics and empiricism has kept them alive.

There are occasional disputes about what counts as science – concerning how best to exclude astrology but include astronomy, about what to say to creationists unhappy with the teaching of evolutionary biology in schools, etc. I leave these disputes to one side here, and begin simply with what are commonly regarded as sciences today. It is widely held that the sciences are not merely knowledge-producing endeavours, but *the* means of knowledge production *par excellence*. Scientific inquiry is our best hope for gaining knowledge of the world, the things that compose it, its structure, its laws, and so on. And the more one investigates, the better it gets. Scientific knowledge is progressive; it renders the natural world with increasing accuracy.

Scientific realism, to a rough, first approximation, is the view that scientific theories correctly describe the nature of a mind-independent world. Outside of philosophy, realism is usually regarded as common sense, but philosophers enjoy subjecting commonplace views to thorough scrutiny, and this one certainly requires it. The main consideration in favour of realism is ancient, but more recently referred to as the 'miracle argument' (or 'no-miracles argument') after the memorable slogan coined by Hilary Putnam (1975, p. 73) that realism 'is the only philosophy that doesn't make the success of science a miracle'. Scientific theories are amazingly successful in that they allow us to predict, manipulate, and participate in worldly phenomena, and the most straightforward explanation of this is that they correctly describe the nature of the world, or something close by. In the absence of this explanation the success afforded by the sciences might well seem miraculous, and, given the choice, one should always choose common sense over miracles.

Some have questioned the need for an explanation of the success of science at all. Bas van Fraassen (1980, pp. 23–5, 34–40), for example, suggests that successful scientific theories are analogous to well-adapted organisms. There is no need to explain the success of organisms, he says.

Only well-adapted organisms survive, just as only well-adapted theories survive, where 'well-adapted' in the latter case means adequate to the tasks to which one puts theories. These tasks are generally thought to include predictions and retrodictions (predictions concerning past phenomena), and perhaps most impressively novel predictions (ones about classes of things or phenomena one has yet to observe). A well-adapted theory is one whose predictions, retrodictions, and novel predictions, if any, are borne out in the course of observation and experimentation. But saying that successful theories are ones that are well-adapted may be tantamount to the tautology that successful theories are successful, which is not saying much. Whatever the merits of the Darwinian analogy for theories generally, one might still wonder why any *given* theory (organism) survives for the time it does, and this may require a more specific consideration of the properties of the theory (organism) in virtue of which it is well adapted. I will return to the contentious issue of the demand for explanations later in this chapter.

The attempt to satisfy the desire for an explanation of scientific success has produced the bulk of the literature on scientific realism. As arguments go, the miracle argument is surprisingly poor, all things considered, and consequently alternatives to realism have flourished. The poverty of the miracle argument and consequent flourishing of rivals to realism stem from difficulties presented by three general issues, which I will mention only briefly:

1 the use of abductive inference, or inference to the best explanation (IBE)
2 the underdetermination of theory choice by data or evidence (UTD)
3 discontinuities in scientific theories over time, yielding a pessimistic induction (PI)

Abduction is a form of inference famous from the writings of Charles Saunders Pierce, inspiring what is now generally called 'inference to the best explanation' (some use the term synonymously with 'abduction' while others, more strictly, distinguish it from Pierce's version). IBE offers the following advice to inference makers: infer the hypothesis that, if true, would provide the best explanation for whatever it is you hope to explain. Note that the miracle argument itself is an abductive argument. Why are scientific theories so successful at making predictions and accounting for empirical data? One answer is that they are true, and this seems, to the realist at any rate, the best explanation. One might even think it the only conceivable explanation, but as we shall see, in light of UTD and PI, this is highly contestable. First, however, let us turn from the particular case of the miracle argument to the merits of IBE as a form of inference in general. There is little doubt that this sort of inferential practice is fundamental to everyday and scientific reasoning. The decision to adopt one theory as opposed to its rivals, for example, is generally a complex process involving many factors, but IBE will most certainly figure at some stage.

Antirealists are quick to point out that in order for an instance of IBE to yield the truth, two conditions must be met. Firstly, one must rank the rival hypotheses under consideration correctly with respect to the likelihood that they are true. Secondly, the truth must be among the hypotheses one is considering. But can one ensure that these conditions are met? Regarding the first, it is difficult to say what features a truth-likely explanation should have. Beyond the minimum criterion of some impressive measure of agreement with outcomes of observation and experiment, possible indicators of good explanations have been widely discussed. Some hold that theories characterized by features such as simplicity, elegance, and unity (with other theories or domains of inquiry) are preferable. Quite apart from the matter of describing what these virtues are, however, and knowing how to compare and prioritize them, it is not immediately obvious that such virtues have anything to do with truth. There is no *a priori* reason, one might argue, to reject the possibility that natural phenomena are rather complex, inelegant, and disjoint. And regarding the second condition for successful IBE, in most cases it is difficult to see how one could know in advance that the true hypothesis is among those considered.[1]

In practice it is often difficult to produce even one theory that explains the empirical data, let alone rivals. This, however, does not diminish the seriousness of the problem. In fact, it turns out that it may be irrelevant whether one ever has a choice to make between rival theories in practice. For some maintain that rival theories are always possible, whether or not one has thought of them, and this is sufficient to raise concerns about IBE. Confidence in the possibility of rivals stems from the underdetermination thesis, or UTD. Its canonical formulation due to Pierre Duhem, later expressed in rather different terms by W. V. O. Quine (hence also called the 'Duhem-Quine thesis'), goes this way. Theoretical hypotheses rarely if ever yield predictions by themselves. Rather, they must be conjoined with auxiliary hypotheses – background theories, related theories, theories

---

[1] A case in which one does have this knowledge is where rival hypotheses are contradictories. See Lipton 1993 for a discussion of this and its implications for IBE.

about the measurement of relevant parameters, etc. – in order to yield predictions. If observation and experimentation produce data that are not as one predicts, one has a choice to make concerning which of the prediction-yielding hypotheses is culpable. One can always preserve a favoured hypothesis at the expense of something else. Since there are different ways of choosing how to account for recalcitrant data, different overall theories or conjunctions of hypotheses may be used to account for the empirical evidence. Thus, in general, there is always more than one overall theory consistent with the data.

In more contemporary discussions, UTD is usually explicated differently. Given a theory, T_1, it is always possible to generate an empirically equivalent but different theory, T_2. T_2 is a theory that makes precisely the same claims regarding observable phenomena as T_1, but differs in other respects. T_2 might, for example, exclude all of the unobservable entities and processes of T_1, or replace some or all of these with others, or simply alter them, but in such a way as to produce exactly the same observable predictions. Given that this sort of manoeuvring is always possible, how does one decide between rival theories so constructed? Here again the realist must find a way to infer to a particular theory at the expense of its rivals, with the various difficulties this engenders.

In addition to challenges concerning IBE and UTD, at least one anti-realist argument aspires to the status of an empirical refutation of realism. PI, or as it is often called, the 'pessimistic meta-induction', can be summarized as follows. Consider the history of scientific theories in any particular domain. From the perspective of the present, most past theories are considered false, strictly speaking. There is evidence of severe discontinuity over time, regarding both the entities and processes described. This evidence makes up a catalogue of instability in the things to which theories refer.[2] By induction based on these past cases, it is likely that present-day theories are also false and will be recognized as such in the future. Realists are generally keen to respond that not even they believe that theories are true *simpliciter*. Scientific theorizing is a complex business, replete with things like approximation, abstraction, and idealization. What is important is that successive theories get better with respect to the truth, coming closer to it over time. It is the progress sciences make in describing nature with increasing accuracy that fuels realism. Good theories, they say, are normally "approximately true", and more so as the

---

[2] Perhaps the most celebrated vision of discontinuity is found in Kuhn 1970/1962. More recent discussions often focus on the formulations of PI given in Laudan 1981.

sciences progress. Giving a precise account of what 'approximate truth' means, however, is no easy task.

So much for common sense. The promise of scientific realism is very much open to debate, and in light of IBE, UTD, and PI, this debate has spawned many positions. Let us take a look at the main players, so as to gain a better understanding of the context of realism.

1.2 A CONCEPTUAL TAXONOMY

Earlier I described realism as the view that scientific theories correctly describe the nature of a mind-independent world. This is shorthand for the various and more nuanced commitments realists tend to make. For example, many add that they are not realists about all theories, just ones that are genuinely successful. The clarification is supplied to dissolve the potential worry that realists must embrace theories that seem artificially successful – those that do not make novel predictions and simply incorporate past empirical data on an *ad hoc* basis, for instance. Realists often say that their position extends only to theories that are sufficiently "mature". Maturity is an admittedly vague notion, meant to convey the idea that a theory has withstood serious testing in application to its domain over some significant period of time, and some correlate the maturity of disciplines more generally with the extent to which their theories make successful, novel predictions.[3] Finally, as I have already mentioned, it is also standard to qualify that which theories are supposed to deliver: it is said that theoretical descriptions may not be true, *per se*, but that they are nearly or approximately true, or at least more so than earlier descriptions.

With these caveats in mind it may be instructive to situate scientific realism in a broader context, as a species of the genus of positions historically described as realisms. Traditionally, 'realism' simply denotes a belief in the reality of something – an existence that does not depend on minds, human or otherwise. Consider an increasingly ambitious sequence of items about which one might be a realist. One could begin with the objects of one's perceptions (goldfish, fishbowls), move on to objects beyond one's sensory abilities to detect (genes, electrons), and further still, beyond the realm of the concrete to the realm of the abstract, to non-spatiotemporal things such as numbers, sets, universals, and propositions. The sort of realist one is, if at all, can be gauged from the sorts of things one

---

[3] See Worrall 1989, pp. 153–4, on the notions of maturity and *ad hoc*ness, Psillos 1999, pp. 105–8, on *ad hoc* theories and novel predictions, and Leplin 1997 on novel predictions.

takes to qualify for mind-independent existence. Though I have just described these commitments as forming a sequence, it should be understood that realism at any given stage does not necessarily entail realism about anything prior to that stage. Some Platonists, for example, appear to hold that ultimately, the only real objects are abstract ones, the Forms, or that the Forms are in some sense "more real" than observables.[4] Scientific realism, in committing to something approaching the truth of scientific theories, makes a commitment to their subject matter: entities and processes involving their interactions, at the level of both the observable and the unobservable. Anything more detailed is a matter for negotiation, and realists have many opposing views beyond this shared, minimal commitment. My own more detailed proposals for realism are outlined in the chapters to come.

I said that 'realism' traditionally denotes a belief in the reality of something, but in the context of scientific realism the term has broader connotations. The most perspicuous way of understanding these aspects is in terms of three lines of inquiry: ontological, semantic, and epistemological. Ontologically, scientific realism is committed to the existence of a mind-independent world or reality. A realist semantics implies that theoretical claims about this reality have truth values, and should be construed literally, whether true or false. I will consider an example of what it might mean to construe claims in a non-literal way momentarily. Finally, the epistemological commitment is to the idea that these theoretical claims give us knowledge of the world. That is, predictively successful (mature, non-*ad hoc*) theories, taken literally as describing the nature of a mind-independent reality are (approximately) true. The things our best scientific theories tell us about entities and processes are decent descriptions of the way the world really is. Henceforth I will use the term 'realism' to refer to this scientific variety only. We are now ready to locate it and various other positions in a conceptual space.

If by 'antirealism' one means any view opposed to realism, many different positions will fit the bill. Exploiting differences in commitments along our three lines of inquiry, one may construct a taxonomy of views discussed in connection with these debates. Table 1.1 lists the most prominent of these, and for each notes how it stands on the existence of a mind-independent world, on whether theoretical statements should be taken literally, and on whether such claims yield knowledge of their putative

---

[4] For a nice summary of the connections between scientific and other realisms, see Kukla 1998, pp. 3–11.

Table 1.1. *Scientific realism and antirealisms*

	The ontological question: mind-independent reality?	The semantic question: theories literally construed?	The epistemological question: knowledge?
Realism	yes	yes	yes
Constructive empiricism	yes	yes	observables: yes unobservables: no
Scepticism	yes	yes	no
Logical positivism/empiricism	yes/no/?	observables: yes unobservables: no	yes
Traditional instrumentalism	yes	observables: yes unobservables: no	observables: yes unobservables: no
Idealism	no	no	yes

subject matter. This is a blunt instrument; an impressive array of viewpoints is not adequately reflected in this simple classificatory scheme, and the reflections present are imprecise. There are many ways, for example, in which to be a sceptic. But the core views sketched in Table 1.1 offer some basic categories for locating families of related commitments.

Traditionally and especially in the early twentieth century, around the time of the birth of modern analytic philosophy, realist positions were contrasted with idealism, according to which there is no world external to and thus independent of the mental. The classic statement of this position is credited to Bishop George Berkeley, for whom reality is constituted by thoughts and ultimately sustained by the mind of God. Idealism need not invoke a deity, though. A phenomenalist, for instance, might be an idealist without appealing to the divine. Given an idealist ontology, it is no surprise that scientific claims cannot be construed literally, since they are not about what they seem to describe at face value, but this of course does not preclude knowledge of a mind-*dependent* reality. As Table 1.1 shows, idealism is the only position considered here to take an unambiguous antirealist stand with respect to ontology.

Instrumentalism is a view shared by a number of positions, all of which have the following contention in common: theories are merely instruments for predicting observable phenomena or systematizing observation reports. Traditional instrumentalism is an even stronger view according to which, furthermore, claims involving unobservable entities and processes have no meaning at all. Such 'theoretical claims', as they are called ('claims about unobservables' is better, I think, since theories describe observables too), do not have truth values. They are not even capable of being true or false; rather, they are mere tools for prediction. In common usage, however, some now employ the term in a weaker sense, to describe views that grant truth values to claims involving unobservables while maintaining that one is not in a position, for whatever reason, to determine what these truth values are. In this latter, weaker sense, constructive empiricism is sometimes described as a form of instrumentalism. And though I have represented instrumentalists in Table 1.1 as subscribing to realism in ontology, some would include those who do not.

Logical positivism, famously associated with the philosophers and scientists of the Vienna Circle, and its later incarnation, logical empiricism, are similar to traditional instrumentalism in having a strict policy regarding the unobservable. But where traditional instrumentalism holds that claims about unobservables are meaningless, logical empiricism assigns meaning to some of these claims by interpreting them non-literally. Rather than taking these claims at face value as describing the things they appear to describe, claims about unobservables are meaningful for logical empiricists if and only if their unobservable terms are linked in an appropriate way to observable terms. The unobservable vocabulary is then treated as nothing more than a shorthand for the observation reports to which they are tied. 'Electron', for example, might be shorthand in some contexts for 'white streak in a cloud chamber', given the path of water droplets one actually sees in a cloud chamber experiment, along what is theoretically described as the trajectory of an electron. It is by means of such 'correspondence rules' or 'bridge principles' that talk of the unobservable realm is interpreted. Given a translation manual of this sort, theories construed non-literally are thought to yield knowledge of the world. The label 'logical positivism / empiricism' covers vast ground, however, and views regarding the ontological status of the world described by science are far from univocal here. Rudolph Carnap (1950), for instance, held that while theories furnish frameworks for systematizing knowledge, ontological questions 'external' to such frameworks are meaningless, or have no cognitive content.

While traditional instrumentalism banishes meaningful talk about unobservables altogether and logical empiricism interprets it non-literally, constructive empiricism, the view advocated by van Fraassen, adopts a realist semantics. The antirealism of this latter position is thus wholly manifested in its epistemology. For the constructive empiricist the observable– unobservable distinction is extremely important, but only in the realm of knowledge, and this feature marks the position as an interesting half-way house between realism and various kinds of scepticism. By scepticism here, I intend any position that agrees with the realist concerning ontology and semantics, but offers epistemic considerations to suggest that one does not have knowledge of the world, or at least that one is not in a position to know that one does. Constructive empiricism goes along with the sceptic part way, denying that one can have knowledge of the unobservable, but also with the realist part way, accepting that one can have knowledge of the observable. (More strictly, constructive empiricism is the view that the aim of science is true claims about observables, not truth more generally, but this is usually interpreted in the way I have suggested.) By adopting a realist semantics, constructive empiricism avoids the semantic difficulties that were in large part responsible for the demise of logical empiricism in the latter half of the twentieth century, and has taken its place as the main rival to realism today.

Table 1.1 does not exhaust the list of "isms" opposed to realism. It does, however, provide a fairly comprehensive list of the reasons and motivations one might have for being an antirealist. For example, the discipline known as the sociology of scientific knowledge is predominantly antirealist. This is not a logical consequence, however, of the desire to study science from a sociological perspective. Sociologists who are antirealists are usually so inclined because of commitments they share with one or more of the antirealist positions outlined in Table 1.1. Though I will not consider this approach to thinking about the sciences in any detail here, it is important to appreciate its influence. Sociological and related methodologies, which attempt to explicate scientific practice and its social, political, and economic relations, both internal and external, represent the major alternative approach to the study of the sciences today, contrasting with the more straightforwardly philosophical approach of realism and constructive empiricism.

Two last positions are worthy of note here, the first of which is actually a family of views belonging to the tradition of pragmatism. This is perhaps the most difficult position to situate with respect to realism, given that most pragmatists would answer 'yes' to all three of the questions posed in Table 1.1, but only some claim to be realists. The difficulty here is that pragmatists adopt a theory of truth that many see as incompatible with realism. For them, truth is an epistemic concept. To say that a statement or theory is true, or that it offers a correct description of the world, is simply to say that it has positive utility – it is useful in some way to believe it. Others hold that truths are what one would believe under epistemically ideal conditions, or in the ideal limit of inquiry. Many realists, however, are uncomfortable with epistemic theories of truth, and adopt instead some version of the correspondence theory, according to which truth is some sort of correspondence between things like theories and the world. But it is doubtful whether one must adopt a correspondence theory to be a realist. There are difficulties associated with explaining what correspondence means, and many prefer to do without. In any case, it does seem that in order to qualify as a realist, one must believe that good theories are reasonably successful in describing the nature of a mind-independent world, but whether this is understood in terms of correspondence truth or in some other way (for example, in terms of a theory of reference or representation) is an open question.

To complete this brief roundup, let me mention what Arthur Fine (1996, pp. 112–50) calls 'NOA', the natural ontological attitude. NOA shares certain motivations with pragmatism, though in addition to rejecting correspondence theories, it rejects all theories of truth including epistemic ones. Its most striking feature is a form of quietism with respect to issues concerning the unobservable that realists and antirealists are wont to contest. As an alternative, NOA prescribes a policy of non-engagement: all ontological claims are on a par, whether about observables or unobservables; beyond merely accepting statements regarding elephants and electrons (as both realists and antirealists do), one should refrain from interpreting such claims by adding that both sorts of objects are real, or that talking about electrons is simply a shorthand for talking about something observable, and so on. NOA rejects both realism *and* antirealism. It is intended as a neutral position for those who find nothing to be gained in debates surrounding them. From the perspective of these debates, however, NOA may seem too anti-philosophical a stand to take. Leaving aside intriguing questions about the potential value and cogency of quietism in this context, I will not consider it further here.

1.3 METAPHYSICS, EMPIRICISM, AND SCIENTIFIC KNOWLEDGE

Armed with a basic summary of realism and its principal rivals, let us turn to the central focus of this work. Earlier I said that much of the controversy surrounding these positions concerns the question of how one should understand scientific claims, in light of the distinctions between the observable and unobservable on one hand, and between the two categories of the unobservable, the detectable and undetectable, on the other. By examining these distinctions one may begin to shed some light on the roles that metaphysics and empiricism play in the interpretation of scientific claims, and the dialectic between them.

The first distinction, between observables and unobservables, concerns things that one can under favourable circumstances perceive with one's unaided senses, and things one cannot. Note that this use of 'observable' and 'unobservable' is different from what is often the case in the sciences themselves. In scientific practice the label 'observable' is usually applied permissively to anything with which one can forge some sort of causal contact, as one does when one uses instruments (such as microscopes) for detection.[5] In the present discussion, however, observables are strictly things one can perceive with the unmediated senses. As Table 1.1 attests, almost everyone thinks one can have knowledge of the observable. This is not to say, however, that interpreting claims about observables is necessarily straightforward. It may be, for example, that the categories of objects and processes one employs to express one's knowledge of the observable are interestingly shaped by the theories one adopts. Indeed, that this is the case for both observables and unobservables is a central tenet of the influential views of Thomas Kuhn (1970/1962), the sociological approaches that followed him, and even some of the logical empiricists who preceded him, who held that "conceptual schemes" shape one's knowledge of the world. I will return to this issue in Chapter 6, but otherwise, for the most part, will take the idea that one has knowledge of observables for granted.

It is the status of the unobservable that has proved most controversial. Logical positivism was, in effect, the founding movement of modern philosophy of science, and the radical empiricism of the positivists has had a lasting impact. It will be useful in what follows to clarify my second distinction, between unobservables that are detectable and those that are not. Let me reserve the word 'detectable' for unobservables one can detect using instruments but not otherwise, and 'undetectable' for those one cannot detect at all (see Figure 1.1). The mitochondrion, for example, is a cellular organelle in which substances are oxidized to produce energy. Though unobservable, one can detect mitochondria using microscopy. A celebrated historical example of a more indirect case of detection is the neutrino, a subatomic particle originally posited by Wolfgang Pauli and theorized about by Enrico Fermi in the 1930s. The neutrino was hypothesized to allow for the conservation of mass-energy and angular

---

[5] See Shapere 1982 for a discussion of the differences between philosophers' and scientists' notions of observation. Shapere examines the conditions under which astrophysicists speak of "observing" solar neutrinos, and also (amazingly) core regions of stars, by means of neutrino detection.

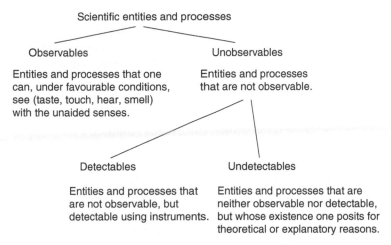

Figure 1.1. Observables and unobservables

momentum in certain subatomic interactions, such as the β-decay of radium-210, and detections of such interactions might thus be viewed as indirect detections of neutrinos. It was not until 1956 that Frederick Reines and Clyde Cowan successfully performed an experiment in which neutrinos were detected more directly. Now consider unobservables whose putative existence cannot be the subject of empirical investigation, whether in practice or in principle. Examples include Newton's conceptions of position and velocity with respect to absolute space, and causally ineffi-cacious entities such as mathematical objects. Even if they exist, such things are undetectable.

Historically, the most pressing challenges to realism have come from those adopting some form of empiricism. This is not to say, however, that all empiricists are antirealists! It may be helpful here to note that empiricism is traditionally associated with two strands of thought which often come together, interwoven. One strand is the idea that sensory experience is the *source* of all knowledge of the world, and this by itself does not preclude an empiricist from being a realist. A realist might accept this first strand while further believing that one can infer the existence of certain unobservables on the basis of the evidence of one's senses. The second strand of empiricism is the idea that all knowledge of the world is *about* experience, and it is this tenet that conflicts with realism, since realists believe claims about things that transcend experience in addition to claims about observables. So an empiricist of the first strand alone may be a realist, but not one of the first and second strands combined. The most adamant critiques of realism stem from those who are committed to the second strand of empiricism, and to violate this commitment is to engage in what its advocates view as a fruitless and misconceived philosophical activity: speculative metaphysics.

What is this metaphysics, then, of which so many empiricists disapprove? To say that there is a conflict between metaphysics and empiricism *simpliciter* is too strong, since many empiricists do metaphysics as it is understood most broadly, as the study of the first or basic principles of philosophy, being *qua* being, and the natures of things that exist. The metaphysics that empiricists disavow concerns the unobservable, and thus any position that endorses speculation of this sort, leading to substantive beliefs about detectables or undetectables, is unacceptable to them. This includes not only speculations about things like universals and causal necessity, which are familiar topics within metaphysics, but also speculations about mitochondria and neutrinos, which are familiar topics within the sciences. But empiricists are generally happy to do metaphysics so long as it does not involve believing speculations about the unobservable. Thus Hume gives an account of causation, not in terms of undetectable necessary connections, but solely in terms of observable events that follow one another. And thus nominalists speak of properties, not as abstract entities like universals, but as sets of observable things to which the predicates associated with these properties apply. The unobservable is likewise an anathema to many empiricist accounts of science. The scientific realist, in maintaining that one can have knowledge of scientific unobservables, engages in the very sort of metaphysical speculation these empiricists reject.[6]

Logical positivism and logical empiricism lost their way, but constructive empiricism has emerged as the main empiricist rival to realism today. Van Fraassen argues for a reconceptualization of empiricism, one of whose goals is to demonstrate the superiority of empiricism over speculative metaphysics. In the remainder of this chapter I will consider his recasting of empiricism, and the question of whether it succeeds in the task of banishing its old adversary. I will argue that it does not, thus opening

---

[6] Van Fraassen (1980, p. 8) defines realism in terms of aspiration: 'Science aims to give us, in its theories, a literally true story of what the world is like; and acceptance of scientific theory involves the belief that it is true.' On this view realism is not necessarily metaphysical, since one might adopt it without endorsing the approximate truth of any claims about unobservables. It seems to me that this is too weak. Realists do believe claims about unobservables, subject to the various caveats I have described, and consequently realism is (on my view) metaphysical.

the door to a detailed consideration of the foundational beliefs of a thoroughly updated scientific realism.

1.4 THE RISE OF STANCE EMPIRICISM

Van Fraassen's reformulation of empiricism occurs within a general framework for thinking about epistemology, the core of which can be described in terms of a tripartite distinction between "levels" of epistemological analysis. At the ground level there are matters of putative fact, or claims about the nature of the world; these are potential objects of belief. Consider, for example, the claim that mammals typically give birth to live young, or that positrons have charge, or that possible worlds exist, or that the only source of knowledge of the world is experience. These are claims about aspects of reality, and if one believes them one takes them to describe these aspects correctly. Factual beliefs do not generate themselves, however. Knowing subjects must acquire them, and when one reflects on how that is done, one arrives at the second level of analysis, the level of stances.

The notion of a stance is intended to be construed rather broadly, but I will use the term to refer to epistemic stances in particular. A stance is a cluster of commitments and strategies for generating factual beliefs. It makes no claim about reality, at least not directly. One might think of them partially, after Paul Teller (2005), as combinations of epistemic "policies" with respect to the methodologies one adopts in order to generate factual beliefs. For example, consider the idea that one should think of explanatory virtue as an important desideratum in determining what to believe, or that one should privilege the methods of the sciences. These are policies regarding the generation of factual beliefs, and policies are not themselves true or false. Certainly, it may be true or false that adopting a particular stance is likely to produce facts as opposed to likely falsehoods, but stances are not themselves propositional for the most part. They furnish guidelines for ways of acting. One does not believe a stance in the way that one believes a fact. Rather, one commits to a stance, or adopts it – they are possible means to realms of possible facts. Crucially, holding a stance is a function of one's *values* as opposed to one's factual beliefs, and though values may be well or ill advised, they are not true or false. (For those critical of the fact–value distinction, it may be possible to speak here simply in terms of different sorts of beliefs.) On van Fraassen's view, as we shall see, metaphysics and empiricism are stances.

The third and final level of epistemological analysis is what I will call the level of meta-stances. Here one finds various attitudes towards the nature of frontline, epistemic stances, and thus ultimately towards the putative facts they generate. One issue at the level of meta-stances is particularly important to the present discussion: the question of which of innumerable possible stances one should adopt. Van Fraassen advocates a view according to which it is rationally permissible to hold any stance and believe any set of facts that meet certain minimal constraints; for example, but not exclusively, those that harbour no logical inconsistency or probabilistic incoherence. This account of rationality, which he calls 'voluntarism', is opposed to the idea that any one stance (and associated set of beliefs) is rationally compelled. I will return to the matter of voluntarism shortly, but first let us come to some understanding of what stance metaphysics and stance empiricism are, precisely.

Earlier I described metaphysical approaches with which empiricists are unhappy as those that endorse speculations about unobservables as a route to belief concerning the unobservable realm. Van Fraassen identifies this with a tradition of analytic metaphysics stretching from seventeenth-century philosophers such as Descartes and Leibniz to contemporary ones such as David Armstrong and David Lewis: 'characterized by the attempted construction of a theory of the world, of the same form as a fundamental science and continuous with (as extension or foundation of) the natural sciences' (2002, p. 231, footnote 1). Henceforth I will simply use the term 'metaphysics' for this sort of speculative approach and 'empiricism' for views that oppose it. The claims of metaphysics annoy the empiricist, but this annoyance is most economically understood at the level of stances. Rather than list the countless factual claims of which empiricists disapprove, one can simply observe that metaphysics is a stance of which empiricists disapprove, which generates annoying factual claims. On van Fraassen's account, stances are generally rich fabrics of interwoven commitments and attitudes, but let me summarize the basic elements of metaphysics very concisely. The core of the metaphysical stance comprises the following epistemic policies:

M1   Accept demands for explanation in terms of things underlying the observable.

M2   Attempt to answer such demands by speculating about the unobservable.

Why should anyone disapprove of these policies? Empiricists hold that via M1, metaphysicians seek to explain things one already understands! Via M2, metaphysicians generate explanantia that are less comprehensible than the explananda with which they begin! These are, it turns out, familiar responses of empiricist philosophers to metaphysics throughout the ages. The empiricist wonders, for example, why she should accept the demand for a deeper explanation of why and how green things form an identifiable group – as she already knows, they are green. And postulating the existence of universals such as greenness, and mysterious relations such as instantiation, is surely more obscure than the fact that some things are green. So argues the empiricist.

Empiricism, conversely, is a stance opposed to the excesses of metaphysics, shared by many historical positions. Again, let me summarize very concisely the core of this position, in terms of the following epistemic policies:

E1   Reject demands for explanation in terms of things underlying the observable.

E2   *A fortiori*, reject attempts to answer such demands by speculating about the unobservable.

E3   Follow, as a model of inquiry, the methods of the sciences.

E1 and E2 are directly opposed to the metaphysical stance.[7] E3, on the other hand, is somewhat puzzling. It is not obvious that the sciences share any particular, substantive, methodological principles, or if they do that they are unique to the empirical stance. Van Fraassen does suggest, however, that one aspect of the sciences of which empiricists approve is a certain tolerance for different beliefs. Scientists routinely disagree, but conflicting beliefs are tolerated and respected as rivals worthy of consideration. One reason he is concerned to portray empiricism as a stance is that he is wary of the charge that, understood as a factual claim, such as 'the only source of knowledge of the world is experience', empiricism may defeat itself. For if empiricism is a factual thesis it will be contrary to other, perhaps metaphysical theses, and though any statement of empiricism would be inconsistent with statements of other views, the principle of tolerance in accordance with E3 demands that one respect contrary factual claims as rivals worthy of consideration. So much for the rejection of metaphysics by empiricists! By ascending to the level of

---

[7] These policies must be qualified if this is to be consistent with van Fraassen's earlier work. There (1980), he distinguishes between belief (taking a theory to be true) and mere acceptance (believing only its observable consequences). Presumably E1 and E2 concern taking explanations to be true, for there may be pragmatic reasons for pursuing metaphysics in some cases. Speculating about unobservables may facilitate the construction of more empirically adequate theories. Without this qualification, there is a tension between E1/E2 and E3, since the methods of the sciences generally favour M1/M2, not E1/E2.

stances, van Fraassen hopes to rid empiricism of any worry of incoherence in its critique of metaphysics.

In any case, E3 is puzzling, not least because a tolerance of contrary factual claims seems too liberal an attitude for the empiricist. Some factual claims are metaphysical, and it is the very business of an empiricist to be intolerant of these claims. Statements about the existence and nature of universals, causal necessity, and possible worlds may be mistaken, but they are putatively factual, and a position that takes such claims as rivals worthy of consideration would be a strange sort of empiricism. Nevertheless, rising to the level of stances does I think help the empiricist to avoid a form of self-defeat. Any plausible definition of empiricism in factual terms, such as 'the only source of knowledge of the world is experience', is likely to make a claim that reaches beyond that which is established in experience. Experience by itself does not rule out the *possibility* of other sources of knowledge. When she defines empiricism as a factual doctrine, the empiricist commits the same sin as the metaphysician: she speculates about the world in such a way as to reach beyond the observable. But this is to engage in metaphysics, and that is why van Fraassen's empiricism cannot be understood as a factual thesis, on pain of defeating itself. One can hardly oppose metaphysics by embracing a metaphysical thesis. The empirical *stance*, conversely, is not part of the metaphysical stance, and to adopt the empirical stance is not to do metaphysics in disguise. Recasting empiricism at the level of stances is thus a means of formulating the position in a way that is not obviously self-defeating.

We are now in a position to ask the question whose answer will determine the very legitimacy of an investigation into the nature of realism. Why should anyone adopt the empirical stance as opposed to its metaphysical counterpart? The reasons had better not make recourse to arguments employing metaphysical premises, or the empiricist will again find herself opposing metaphysics by doing metaphysics. And thus we find ourselves with two stances, the empirical and the metaphysical, and wanting an argument for why the former is preferable to the latter. What, then, is the case against metaphysics?

1.5 THE FALL OF THE CRITIQUE OF METAPHYSICS

I submit that there can be no case against metaphysics, or more correctly, no case for a fair-minded, non-dogmatic metaphysician to address. To understand why this is so, one must engage a specific concern at the level of meta-stances: identifying an appropriate criterion or criteria with which to facilitate choosing a stance. Van Fraassen suggests two criteria: one that is uniformly applicable to anyone's choice of stance, and another whose application varies across stance holders. The uniform criterion is rationality. One should adopt a stance that is rational and reject those that are not. The variable criterion is the set of values that leads an agent to adopt one stance over another.

I will return to the issue of values momentarily, but first let us consider van Fraassen's conception of rationality, which is famously thin. It is rationally permissible, he says, to hold any stance or believe any set of facts that is logically consistent and probabilistically coherent. Incoherence was originally explicated (1989) in terms of holding combinations of beliefs that are exploitable by Dutch books to the detriment of the belief-holder (making bets all of whose possible outcomes are unfavourable), and consistency and coherence are usually understood as logical constraints, straightforwardly applicable to propositional things like factual beliefs. Stances, however, are in large part non-propositional, so in this context mere *logical* consistency and coherence will not suffice. At least part of what is intended by incoherence here must have a pragmatic dimension, and indeed, van Fraassen (2005, p. 184) holds that the 'defining hallmark' of irrationality more generally is 'self-sabotage by one's own lights'. Self-sabotage is broad enough to include such unfortunate circumstances as believing contradictions and probabilistically incoherent combinations, as one might do on the level of facts, but it may also include circumstances in which the stance one adopts has pragmatic failings, such as a combination of attitudes or policies that tend to undermine or conflict with one another. Note that on this view, different and mutually incompatible stances may be rational – no one stance and resultant set of beliefs are compelled. Van Fraassen calls this meta-stance 'voluntarism'.

Let us now return to values. Recall that in addition to rationality, agents' values furnish criteria for their choice of stance. If one's values promote a commitment to the empirical stance, one will reject metaphysics. After all, E1 and E2 are directly opposed to M1 and M2. The empiricist rejects metaphysics by committing to epistemic policies that are incompatible with it. But does this offer a case against metaphysics? To the consternation of the empiricist, it does not. For if rationality is the only constraint that applies uniformly to all agents adopting stances, and different, mutually incompatible stances are rational, then the framework for debate on the level of stances is relativistic. Relativism is premised on the idea that there is no view from nowhere, no view that cuts across perspectives so as to serve as a sufficient common ground from which to debate. If it turns out that metaphysics is rational, empiricists may nevertheless claim that it is wrong-headed from their perspective. The qualifying phrase 'from one's perspective', however, is inseparable from any statement of the correctness of adopting a stance. Saying that different communities have different values is shorthand for saying that correctness and incorrectness are relativized to perspectives, and have no meaning otherwise.

Comparing M1 and M2 to E1 and E2, one finds different policies supported by different intuitions, or values, concerning two things: what needs explaining; and what counts as obscure or unilluminating. Many criticisms of stances that meet the constraint of rationality are cogent only from within the confines of some other stance, and this cogency is not preserved "outside". Thus, if empiricists hope to offer a case against metaphysics that is telling for the metaphysician, not merely for someone who adopts empiricist values that metaphysicians need not share, they must demonstrate the *irrationality* of metaphysics, because rationality is the only stance-transcendent criterion for choosing a stance. In other words, the empiricist must show that metaphysics sabotages itself, or more specifically, that if one adopts the epistemic policies of metaphysics, there are derivable consequences of which even metaphysicians would disapprove.

The task, then, is to demonstrate that metaphysics fails by its own lights, but how? Perhaps one could argue that the factual claims of metaphysics are problematic. Van Fraassen (1989) himself argues, for example, that the concept of a law of nature is incoherent. But even if it turned out that *every* current metaphysical concept was incoherent, this would not amount to a demonstration of the irrationality of metaphysics. One interesting consequence of understanding metaphysics and empiricism as stances is that they are not (exclusively) identifiable with any one set of factual beliefs. Stances underdetermine the factual beliefs they produce. Over philosophical time, both metaphysics and empiricism have survived many changes in the beliefs with which they are associated and no doubt will again. For this reason, van Fraassen (2002, p. 62) is clear that stances are not identical to the factual claims with which they may be associated at any given time. Thus, no demonstration of the irrationality of believing such factual claims can entail the irrationality of adopting a stance.

Let us focus, then, on the stance itself. Perhaps there are commitments, standards, or principles accepted by metaphysicians that the metaphysical stance itself fails to meet or exemplify. If so, this would constitute the sort of pragmatic incoherence the empiricist requires in order to demonstrate that metaphysics is irrational. There are suggestions to this effect throughout van Fraassen's critique. Let me summarize the relevant principles as follows:

P1   No form of inquiry into the nature of the world should be immune to the possibility of error, or failure.

P2   Correct logical or grammatical form should not be considered sufficient to render claims about the world substantive.

P3   The epistemic status of one's criteria for theory choice should be linked to the epistemic status of one's theories.

It seems reasonable that both metaphysicians and empiricists should accept P1–P3, so let us examine each in turn, and consider why one might think metaphysics fails to satisfy them.

First, consider P1. Van Fraassen and empiricists generally are sometimes heard to complain that metaphysics has the character of a particularly futile game. Its futility is evidenced by the fact that no one ever wins or loses, and perhaps most damagingly, it never ends! If some part of metaphysics is shown to be inconsistent, it simply reinvents itself. One always has the option, it seems, of retreating to another position within the game of metaphysics that is immune to the criticism applied, and this violates P1, the idea that no form of inquiry should have this kind of immunity. It would not be difficult, I suspect, to find some measure of sympathy for this complaint among those who are interested in the sciences. A great deal of speculation in metaphysics is too far removed from the sciences to generate much interest or care on the part of realists, for instance, at least in the context of thinking about scientific knowledge. This, however, merely expresses a taste, and expressions of taste are not demonstrations of irrationality. Metaphysicians should accept P1, since metaphysics is fallible, but one must take care not to conflate metaphysical claims and theories with the metaphysical stance itself, any more than one would conflate the empirical stance with any particular empirical claim or theory. When metaphysical claims are found to be problematic, one tries something else. Clearly, then, particular theories *can* lose out, and it is not a pointless game after all. It is in the nature of the stances that generate these candidates for knowledge, however, to go on. Thus it seems that P1 is no threat to the metaphysical stance.

Consider P2. Van Fraassen challenges metaphysicians to show that their claims are substantive. They should amount to more, he says, than 'coherent nonsense'. Merely correct logical or grammatical form is insufficient to demonstrate that metaphysical claims exemplify reasonable attempts to say something substantive about the world. Again, I suspect that metaphysicians would agree with P2, but it is an odd sort of thing to be asked to prove the substance of one's claims, especially in the context of one's own inquiry. In response to the question of how anyone could think that M1 and M2 lead to substantive contentions, one might legitimately wonder what sort of answer would suffice. There is an interesting question here of the burden of proof. In just the same way that the empiricist wants to know what reason anyone might have for thinking that metaphysical claims are substantive, the sceptic might well ask constructive empiricists to show, for example, that their claims about the world are, in fact, something more than coherent nonsense, and so on and so forth. Perhaps only the solipsist of the present moment is safe from this line of questioning. At the end of the day, the only thing anyone can do in response to this sort of question is to point to his or her own epistemic practices, and the values that favour them, and this takes us to P3.

Metaphysical theories, says van Fraassen, are evaluated in terms of purely subjective values and probabilities of success. These values, however, such as preferences for theories that maximize simplicity, scope, or explanatory power, are not *epistemic* values. That is, they are not linked to truth, or at least one has no reason to think they are. Metaphysicians thus suffer from a form of 'false consciousness': they apply their subjective values and probabilities of success in pursuit of truths, but there is no reason to think that such application leads to anything other than theories they like. I submit, however, that van Fraassen is not in a position to make this charge, given his voluntarism. Once again, it seems reasonable that metaphysicians should accept P3, but they disagree with the empiricist's evaluation of the epistemic status of their criteria for theory choice. Consider the case of scientific theories and their epistemic status. Under certain conditions, realists think it is reasonable to infer the approximate truth of our best theories involving unobservables, and their criteria for theory choice include such things as maximizing simplicity, scope, explanatory power, etc. Empiricists demur. These criteria are at best indicative of truths about observables, they say. But does this disagreement entail that at least one of these parties is being irrational? It is hard to see how it could – neither position is rationally compelled, and neither, it seems, is guilty of inconsistency or incoherence.

Both metaphysicians and empiricists make a leap from what is strictly entailed by the observable data, as a matter of faith, perhaps, but in different ways, consistent with the values to which they subscribe. It should thus be clear that P3 is no threat to metaphysics. It is precisely *because* metaphysicians think their criteria for theory choice are epistemically significant, as a result either of a voluntaristic choice or of reasons to be adduced, that they believe our theories might well be close to the truth. There is no pragmatic incoherence in this. There is, no doubt, a difference in degree between the speculation about unobservables that is most commonly part of realism, and much of what takes place in metaphysics more generally. In both cases, speculations about unobservable entities and processes are intended to be consistent with the observable data, but often scientific theories seem to take a greater risk, because they often make *novel* predictions and other metaphysical theories do not. Not all sciences make novel predictions, however, and differences in risk are differences in degree, not kind. There is no rationally compelled answer to the question of how much is required in order to make a form of inquiry acceptable. Where one draws the line here will depend on the values one has, not on matters of rationality.

In concluding this chapter I believe we are now in a position to appreciate why the realist cannot be arrested by the empiricist critique of metaphysics. The critique is subject to a form of relativism that renders it effective only to the ears of empiricists. It appeals to values and policies that empiricists share, but that need not be shared by other rational agents. Only if it could be demonstrated that the metaphysical stance is incoherent by its own lights would the empiricist have a critique that escapes this conundrum, but this is asking too much. At one point, van Fraassen (2000, p. 277) characterizes what it is to be rational in terms that I think, despite his deep commitment to empiricism, embrace the metaphysical stance:

Nothing more than staying within the bounds of reason is needed for [the] status of rationality. Not good reasons, not a rationale, not support of any special sort ... nothing is needed above and beyond coherence. Thus any truly coherent position is rational.

On his conception of epistemology the threshold for rationality is low, and as a consequence the threshold of irrationality is very high indeed. When the sceptic challenges the constructive empiricist to prove that it is not irrational to believe the observable content of our best theories, I do not think the latter has much to answer for. Empiricists choose forms of inquiry that fit with their values, epistemic and otherwise, and some of these tell them the sceptic's life is not worth living. The same applies to the metaphysician. One may decide, in accordance with one's values, what forms of inquiry to pursue. That is our prerogative, after all. But few if any prerogatives transcend all possible stances, and there can be no radical critique of metaphysics by empiricism.

The metaphysics of the sciences concerns the observable and unobservable parts of the world described by scientific theories, both explicitly and implicitly. The epistemology of the sciences concerns the specific methods used to generate scientific claims, the justification or confirmation of these claims, whether they constitute knowledge, and if so, what sort. The influence of logical positivism during the birth of the philosophy of science as a separate discipline in the late nineteenth century, and throughout most of the twentieth century, led to a vestigial neglect of metaphysical questions in connection with realism. Those investigating problems such as the nature of causation, laws of nature, and conceptions of natural kinds have done so largely in isolation from debates between realists and antirealists. Metaphysical issues have been the purview of the philosophy of particular sciences: space and time, evolutionary biology, quantum mechanics, and so on. The neglect of metaphysics in the context of realism, however, is a mistake. For there is a sense in which the metaphysics of science is a precursor to its epistemology. One cannot fully appreciate what it might mean to be a realist until one has a clear picture of what one is being invited to be a realist about.

In the further chapters of this book, I will propose an answer to the question of how to construe realism by developing a metaphysics that underpins it. The aim of this endeavour is an integrated account of the unobservables of which scientists speak in detail, like mitochondria and neutrinos, and those features of reality that realists sometimes take for granted but say little about, like causation and laws. The result, I hope, is a reunion of arguments about the natures of things in the world with those about how one can know these things – a reunion that redresses the separation of metaphysics and epistemology in the context of scientific knowledge.

CHAPTER 2

Selective scepticism: entity realism, structural realism, semirealism

2.1 THE ENTITIES ARE NOT ALONE

Scientific realists invite questions about their metaphysical beliefs, often perhaps unwittingly. In their accounts of scientific knowledge, they routinely invoke not only unobservable entities and processes commonly discussed by scientists, but also things whose natures generally fall outside the remit of the sciences, such as causation, laws of nature, and the idea that scientific taxonomies divide the world into natural categories, or kinds. The recourse to these latter metaphysical notions in support of realism is not problematic *per se*, but a lack of attention given to spelling them out can have problematic consequences. While these topics are central to metaphysics and many realists investigate them, few offer unified accounts in connection with specific proposals for realism. In the absence of such details, the views of realists are sometimes associated by default with the metaphysical speculations of great, systematic philosophers of the past, from ancient and medieval, up to and including early modern times. Unfortunately for the realist, some of these speculations are outmoded today, especially in a modern scientific context. In this chapter I will begin the process of spelling out in a more detailed way what I believe scientific realism has become, and thereby initiate a proposal for its metaphysical foundations.

Chapter 1 began with a rough, first-approximation definition of realism: scientific theories correctly describe the nature of a mind-independent world. This first approximation, however, is naïve in several respects, and this leaves it open to several immediate objections. In order to remedy this situation a number of qualifications are usually made. These include the idea that realists should only commit to theories that are genuinely successful and not merely *ad hoc*, as evidenced by the nature of their predictions, retrodictions, and novel predictions. Another restriction is to theories that are sufficiently mature. A mature theory is one that has survived for a significant period of time and has been rigorously tested, perhaps belonging to a discipline whose theories typically make novel predictions. A further qualification concerns the degrees of accuracy to which theories describe the world. Realists accept that often theories are not true, but nevertheless hold that mature theories are typically close to the truth and increasingly so, within a discipline, over time.

These nuances by no means render realism impervious to antirealist scepticism, though. Taking them into account, one may yet worry about inference to the best explanation (IBE), the underdetermination of theory by data (UTD), and the pessimistic induction (PI). For present purposes, however, it will serve to distinguish IBE and UTD on one hand from PI on the other. There are two important ways in which concerns about IBE and UTD differ from those about PI. The first is that the former are challenges that face *any* account of realism, in principle, because abductive inferences play an inextricable role in many contexts of scientific reasoning, and if UTD is a genuine problem, its challenge can be formulated in connection with any realist conception of scientific theories or claims. PI, on the other hand, may not apply to all forms of realism, and this is linked to the second difference between these antirealist arguments. Given the generic nature of concerns about IBE and UTD, debates regarding these challenges have had little influence on the evolution of realism as a philosophical position. But since the degree to which PI is worrying may vary, depending on the precise specification of one's realist commitments, this and related concerns have played a dramatic role in shaping the modern face of realism. The goal of this chapter is to consider the impact of these worries, and to arrive at a preliminary sketch of a proposal for how one might think of realism now.

None of the disciplines and research specialties of the sciences have been immune to change throughout their histories. Many of the best theories of our scientific past, compelling and popular in their day, are now regarded as profoundly mistaken. Many terms for unobservable entities and processes we once thought referred to things in the world have since been rejected as non-referring. We no longer believe, for example, in the crystalline spheres on whose surfaces extra-terrestrial bodies travel the heavens, according to ancient astronomy, or that combustion is a process in which materials exude a substance called 'phlogiston', as was held in the late seventeenth and eighteenth centuries, or that heat is a conserved fluid called 'caloric', as many thought in the late eighteenth and early nineteenth centuries. This sort of discontinuity is something the realist must come to terms with. Indeed, PI argues that, given such track records of discontinuity, one has grounds for an induction, the conclusion of which is that current theories are also likely false and make reference to entities and processes that one will come to regard as fictions at some future stage of scientific development. Surely the position of the realist is untenable, given that theories are revised and replaced over time. Realism, says the antirealist, is not suited to theoretical instability.

There are several possible responses to PI. Some are happy merely to say that although many past theories are now considered false and contain non-referring terms, realism is tenable so long as successions of theories within disciplines are increasingly approximately true. Others contend that the pool of data serving as the ground of the induction is too small to permit a credible conclusion. Many responses to Larry Laudan's (1981) version of PI, for example, take issue with the list of theories he cites as evidence, arguing that if one factors in the further qualifications assumed by realists – the notion of maturity, the exclusion of *ad hoc* theories, the importance of novel predictions, and so on – the data for pessimism are greatly reduced. No doubt, if it is the case that radical discontinuities in scientific theorizing are relatively rare, the realist is in better shape than PI suggests. But I will not pursue these arguments here. I will return to the issue of approximate truth in Chapter 8, but determining the proportion of past theories one would now regard as significantly alien to currently accepted theories would require exhaustive case studies of the history of science, and such a gargantuan task is beyond the scope of this (and probably any) work. It is also fraught with difficult historical questions: which past theories commanded sufficient support to merit inclusion in the survey, and how is the relative seriousness of degrees of discontinuity to be judged? Here I will simply proceed with the reasonable assumption that significant discontinuity is a fact of much of scientific history, and the realist should have something to say about it.

The most promising suggestion for realism here comes from a familiar adage. As in life generally, so too in science: do not believe everything you are told. Not all aspects of scientific theories are to be believed. Theories can be interpreted as making many claims about the nature of reality, but at best one has good grounds, or epistemic warrant, for believing some of these claims. Only some aspects of theories are likely to be retained as the sciences march on. I will refer to any approach that takes this advice seriously as a form of *selective scepticism*. The primary motivation for this modification to realism *simpliciter* is to pick out, from among the numerous claims embedded in theories, the ones that are most epistemically secure and thus likely to survive over time. On this view, the question of how best to be a realist boils down to the question of which aspects of theories one should believe. It will come as no surprise, however, that different philosophers have drawn the line between what one should and should not believe in very different places! Several of these insights are on the right track, but it also seems clear to me that none has drawn the line in precisely the right place. Perhaps by considering where these important, previous attempts have gone both right and wrong, one may illuminate a promising path for realism.

Traditionally, selective sceptics among realists have fallen into two broadly defined camps: entity realism, and structural realism.[1] I will begin with entity realism (ER), which is perhaps the best-known and thus most influential approach to selective scepticism. ER is the view that under certain conditions, one has good reason to believe that the entities described by scientific theories exist in a mind-independent reality. It is this aspect of theory – claims about the existence of specific entities – that one can reasonably believe to be true. This is the positive thesis of ER, but it also has a negative aspect. While endorsing certain existential claims, ER is generally sceptical about the theories in which, *inter alia*, these entities are described. It is this combination of positive and negative aspects that allows ER to function both as a form of realism, and as a form of selective scepticism hopefully capable of offering a response to PI. It is a form of realism because it endorses a knowledge of unobservable entities, yet by espousing an antirealism about theories more generally, it may happily accept that much of the further contents of theories has changed over time and may change again in future.

What, then, are the conditions under which one is entitled to believe in entities, according to ER? Different advocates of the position say slightly different things here, but all point to the significance of our *causal* contact with the entities involved. Ian Hacking (1983) and Ronald Giere (1988) argue that in practice, scientific experimentation routinely implies the existence of certain entities. When one can manipulate them so as to intervene in other things, says Hacking, effectively using them as tools for scientific investigation by exploiting their causal powers, one cannot doubt their existence. As an example of this sort of exploitation, he discusses the use of electrons to study weak neutral current interactions among sub-atomic particles. The epistemic virtue of causal contact is also championed by Nancy Cartwright (1983, Essay 5), who argues that when one accepts a

---

[1] Psillos 1999, ch. 7, is suspicious of this dichotomy. In what follows, I present further but different reasons for suspicion.

causal explanation of a given phenomenon, one must accept the reality of the relevant cause. If one believes that one can use a laser to ionize an atom (knock out one or more of its electrons, leaving it with a positive electrostatic charge), one must believe in the reality of the cause, viz., lased light. In general though in different ways, entity realists appeal to the epistemic significance of our causal connections to particular entities.

The pros and cons of ER have been much discussed, but keeping in mind the present goal of charting the evolution of realism, let me simply note its best feature and two rather telling objections. The good news is that there is considerable evidence to support the idea that when one manages to forge significant causal contact with entities, they are retained when theories involving them change over time. Numerous theories about the nature of the electron, for example, have come and gone since J. J. Thomson speculated that the 'cathode rays' he was experimenting on in 1897 might be composed of a stream of 'corpuscles', but the entity itself still has a place in current theory. There is a *prima facie* case, it seems, that ER may be a refuge for the realist in the face of historical discontinuity.

Two serious objections, however, put this in doubt. The first concerns the precise location of the line ER draws between that which realists are advised to endorse, and that about which they are advised to remain sceptical. ER endorses claims about the existence of certain entities and is sceptical about theories generally, but although existential claims about entities and other theoretical claims are undoubtedly different things, they are not so easily separated when it comes to knowledge. One cannot have knowledge of the existence of entities in isolation. In order to know that something unobservable exists, one must know the details of at least some of its *relations* to other things – relations, for example, to instruments of detection, or to instruments of manipulation and the aspects of phenomena in which one hopes to intervene by manipulating the entities in question. Entities are capable of these relations because of the properties they have, and properties and relations are precisely what theories describe, so by asking the realist to believe only in the existence of certain entities but not further aspects of theories, ER asks too much. It asks the realist to endorse existential claims, but to be sceptical of the very knowledge that gives the realist reason to endorse these claims, and that is difficult advice to accept. In order to be a realist about entities, one must be a realist about at least some aspects of theory also.[2]

---

[2] Hacking 1982 seems to allow knowledge of a limited number of 'home truths' or 'low-level' generalizations about manipulated entities, which may take him beyond the strict terms in which

A second worry about ER concerns the response it yields in the face of PI. In cases where, according to ER, one's causal knowledge of an entity is sufficient to merit an existential claim, one is ostensibly safe from worries about discontinuity. Although theories regarding the nature of an entity may change over time, one may carry one's knowledge of its existence throughout. There is a sense in which this is no doubt the case. If indeed, starting with Thomson, Robert Millikan, Ernest Rutherford, and throughout the twentieth century, a long line of experimentalists interacted with the same entity, there is a sense in which it is fair to say that all are talking about the same thing – they all referred to the electron. (This assumes some version of the causal theory of reference, itself controversial; I will consider this worry in section 2.5.) Perhaps not all of these experimenters had sufficient causal knowledge of the electron to satisfy ER – not all manipulated it, for instance – but no doubt many met the required standard. There is another sense, however, in which it is too glib to say that all are talking about the same thing, because over time, scientists have believed extraordinarily different things *about* electrons. Given discontinuity, successful reference alone is small comfort if it turns out that different generations of scientists had radically different conceptions of the properties of electrons. The problem here is that ER is too crude. It endorses existential statements, but what the realist requires is something more refined: a knowledge of the specific properties and relations on which existential claims are based, and that are likely to survive over time.

The remainder of this chapter seeks to expose such refinement in the further evolution of scientific realism. In anticipation, let us gather some morals from the preceding discussion of ER. I have argued that it gives the wrong diagnosis of where the dividing line is between what a realist should and should not believe. I have also suggested that it gives unsatisfactory answers to important historical questions, concerning what realists should believe about entities over time. Despite these (I think) fatal problems, however, I believe there is something important to be learned from ER, and this will turn out to be crucial in coping with worries like PI. Why should anyone believe in the existence of unobservable entities? Because, in some cases, one is connected to them causally. Belief in the existence of scientific entities is rarely black and white. One does not divide unobservables into two categories: those one is certain exist, and those one is certain do not.

I have described ER. Morrison 1990 argues that ER cannot restrict knowledge as Hacking suggests: in order to produce the causal processes by which entities are known, one requires further theoretical knowledge.

In the sciences one finds graded spectra of commitment, and the realist must emulate this. There are some things about which one is quite certain, as a consequence of impressive abilities to exploit their causal powers in intricate and fantastic ways. Where causal contact is more attenuated, one is appropriately less confident. At the far end of any given spectrum are entities one is relatively unsure about – the subjects of relatively indirect detections or speculations about undetectables. (Further still, perhaps, are fictitious entities, knowingly countenanced to play merely instrumental or heuristic roles.) But ER gives us a clue about what sorts of things realists can believe in. I will return to this thought. First, however, let us consider the second main realist proposal for selective scepticism.

2.2 LESSONS FROM EPISTEMIC STRUCTURALISM

Structural realism (SR) is the view that insofar as (mature, non-*ad hoc*, etc.) scientific theories offer approximately true descriptions of a mind-independent reality, they do not tell us about its *nature*, or more specifically, the nature of its unobservable parts. Rather, they tell us about its *structure*. What this could mean, however, is a matter of some controversy. I will consider the likely possibilities momentarily. Before delving into the details, however, note that SR, like ER, has both positive and negative aspects, and it is this union that allows it to function as a form of selective scepticism. While endorsing claims about certain structures, SR is generally sceptical about the natures of the entities that might be thought to inhabit these structures. It is a form of realism because it endorses a knowledge of the structure of the unobservable, but is sceptical in its antirealism about substantive claims regarding entities. Contemporary proponents of SR thus see in the position a safe route between opposed forces. On the one hand, if theories are to some extent correct in mapping the structures of the natural world, one has an explanation for the success they afford in allowing us to predict and manipulate natural phenomena. The miracle argument for realism is thus accommodated. On the other hand, if one takes only certain parts of theories consisting in descriptions of structures to describe the world, one ostensibly has a ready response to the worry of theoretical discontinuity over time. One is in a position to sacrifice anything but the desired structural aspects of theories to PI.

Realist advocates of SR take inspiration from predecessors who are by no means unified in their philosophical motivations, including Henri Poincaré, Bertrand Russell, Ernst Cassirer, Moritz Schlick, and Rudolph Carnap (Gower 2000 gives a historical survey of some early structuralists).

As a consequence, discussions of SR are susceptible to conflations of different notions of structure, and often there is significant ambiguity regarding these notions to begin with. In order to clarify the important role of SR in the evolution of realism I will focus on the concept of structure in particular, for by doing so, I believe one learns valuable lessons from previous, problematic, structuralist accounts. The most plausible face of realism today, I will suggest, is indeed a form of structuralism, but one that pays careful attention to the morals of ER.

What then is intended by the idea of structure, here? Let us begin with the ordinary, everyday concept of structure. If asked to describe the structure of the picture frame on my office wall (containing the photograph of my favourite armillary sphere), I would list the parts of the frame – the glass, the sides, the back, the wire brace – and describe how they are related to one another – geometrically, by glue, etc. Informally, the idea of a structure has to do with relations between the elements of some system of elements. Structuralism focuses on the relations themselves rather than on the things standing in these relations, the relata. In the contemporary literature, SR comes in two flavours: *epistemic* SR, and *ontic* SR.[3] Epistemic versions place a restriction on scientific knowledge; proponents hold that one can know structural aspects of reality, but nothing about the natures of those things whose relations define structures in the first place. The natures of the entities are beyond the proper grasp of our quest for knowledge. Ontic versions, more radically, do away with entities altogether; proponents hold that at best we have knowledge of structural aspects of reality, because there is in fact nothing else to know.

I will contend that the most reasonable form of SR is *both* epistemic *and* ontic, but in ways different from what is suggested by current proponents of epistemic and ontic SR. In particular, the putative distinction between a knowledge of the structure of reality and a knowledge of its nature is difficult to maintain and profitably collapsed. In what follows, I will briefly review the epistemic tradition of SR *en route* to offering what I take to be a more promising proposal for a non-naïve or sophisticated realism. Epistemic SR faces fatal difficulties, I believe, but gives genuine insight into the promise of selective scepticism as a strategy for the realist. I will leave a detailed consideration of ontic SR to Chapter 3.

---

[3] The distinction is due to Ladyman 1998. The same distinction is found in Psillos 2001 under the labels 'restrictive' and 'eliminative' SR, respectively. For a defence of epistemic SR, see Worrall 1989 and 1994, Zahar 1996, and Worrall and Zahar 2001. Ontic SR is favoured by Ladyman 1998, French 1998 and 1999, and French and Ladyman 2003.

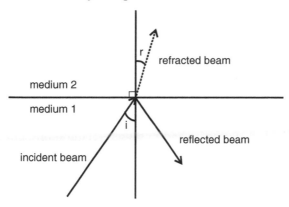

Figure 2.1. Incident, reflected, and refracted light beams
at the interface of two media

John Worrall (1989) brought epistemic SR back into the philosophy of science after a significant hiatus, identifying Poincaré as its founding father. Poincaré, however, is not very informative on the subject of what structures are, precisely. His most striking case study, from *Science and Hypothesis* (1952/1905, ch.10), is the transition in theories of light from the wave optics of Augustin Fresnel in the early nineteenth century to the electromagnetic theory of James Clerk Maxwell. Among his many achievements, Fresnel developed a set of equations relating the intensities of incident, reflected, and refracted light when a beam passes from one medium into another having a different optical density (see Figure 2.1). In opposition to the Newtonian, corpuscular theories of light that preceded him, according to which light is understood as a stream of particles, Fresnel believed that light is a wave-type disturbance in the ether, a mechanical medium. Wave forms, he thought quite reasonably, are disturbances *in* something that propagates them. Just as water waves are disturbances in water, light waves must be disturbances in another sort of medium, the luminiferous ether – or so many thought during the course of the nineteenth century.

Maxwell's theory of electromagnetism later incorporated visible light as one of the many forms of electromagnetic radiation. Though Maxwell attempted to formulate his theory in such a way as to make it consistent with the existence of a luminiferous ether, however, the failure of experimental results to detect the ether towards the end of the nineteenth century hastened its demise. Ultimately, Maxwell's theory was accepted, not as a description of ethereal disturbances, but simply as a description of oscillating electromagnetic field vectors. Nevertheless (and this is where structuralists stop and take note, triumphantly) Fresnel's equations survive intact in Maxwell's theory! Precisely the same equations can be derived from Maxwell's later account of electromagnetism. The moral, says Worrall: despite changing views concerning the nature of light, here a description of the structure of light survives from one theory to the next. The structure of light, he claims, is given by the mathematical equations common to Fresnel's and Maxwell's theories.

The case study is suggestive and raises several questions, but let us begin with the most basic. What does it mean to say that these mathematical equations are indicative of structure? It is not immediately apparent how this is to be understood. Certainly, it is insufficient for the realist simply to point at the equations of theories and claim that they describe reality, since constructive empiricists, instrumentalists, logical positivists, and idealists would say the same, subject to their own interpretations. Without further clarification, the ambiguity of the appeal to equations renders it too weak to amount to a statement of realism. For assistance on this point, epistemic structuralists have turned to Russell, who developed his own structuralist epistemology in a rather different context. Russell believes that the only 'knowledge by acquaintance' one has is of one's own sense data – mental objects or events, formed in momentary experiences or perceptions. The external world is not known by acquaintance, but 'by description', and these descriptions are limited in the information they convey. According to Russell, the structure of one's sense data mirrors the structure of the world, and that is all that can be known of it. 'No non-mathematical properties of the physical world can be inferred from perception' (1927, p. 253).

It is clear straight away that Russell's project in epistemology differs from that of the contemporary scientific realist. Unlike Russell, realists generally accept that one has more than merely structural knowledge of external observables, for example. Nevertheless, insofar as epistemic SR is interested in limiting knowledge of the unobservable realm to its structure, Russell's approach is a potentially useful model. For one may employ his understanding of structure *whatever* one's view of where to draw the line between things one can know only structurally, and things one can know more fully. In fact, Russell's definition of structural identity is a familiar one, commonly found in mathematics and parts of physics. This is how it goes: 'We shall say that a class α ordered by the relation R has the same structure as a class β ordered by the relation S, if to every term in α some one term in β corresponds, and vice versa, and if when two terms in α have the relation R, then the corresponding terms in β have the relation S, and vice versa' (1948, p. 271; see also 1927).

The crucial point to note about this conception of structural identity is that the members of α and β, and the relations R and S, need not bear any qualitative similarity to one another. That is, the members of α may be completely different kinds of things from the members of β, and R and S may be completely different sorts of relations. The only requirement is that R and S have a purely formal similarity. Structure is understood here as a property *of* relations – a higher-order, formal (logical or mathematical) property. For example, consider an assortment of delicious cakes of different masses. The relation 'heavier than' is a qualitative relation between various cakes that obtains in virtue of their masses, which are first-order properties of the cakes. Similarly, imagine a collection of spoons of different lengths. The relation 'longer than' is also a qualitative relation. The heavier-than and longer-than relations have something in common: the higher-order property of a total ordering. Thus, one might imagine a set of cakes and a set of spoons, with their respective masses and lengths and heavier-than and longer-than relations, that satisfies Russell's definition of structural identity. But crucially, on Russell's view, no qualitative, first-order properties or relations of the objects need be known. So long as there *is* a relation (of any sort) between some elements of a system, that is sufficient to determine a structure, and this has generated controversy regarding the application of this concept to the epistemic context of realism.

William Demopoulos and Michael Friedman (1985, pp. 628–9) argue that by suggesting only formal properties of the world can be known, Russell invites a fatal objection. They cite M. H. A. Newman's (1928, p. 140) criticism that on Russell's definition, the claim that a system has a particular structure tells us nothing about it other than its cardinality, because *any* collection of elements can be arranged so as to exemplify a given structure so long as there are enough of them. Given any set α and any arbitrary structure W, it is a consequence of set theory or second-order logic that there exists some relation in α having structure W, so long as W is compatible with the number of elements in α. (The relevant theorem states that every set α determines a full structure, viz. one that contains every relation of every arity on α; this forms the basis for a model of the language of higher-order logic.) Thus, on Russell's approach, the claim that some aspect of the world has a particular structure is trivially satisfied. One knows that such claims are true, subject to cardinality, before one even begins an empirical investigation into the unobservables in question.

Epistemic structuralists take further inspiration from Grover Maxwell, who married Russell's approach to the Ramsey sentence to produce his own version of SR. To obtain the Ramsey sentence of a theory one conjoins its axioms, and then replaces all of the unobservable terms with existentially quantified predicate variables. An unobservable entity whose place is held by a variable in a Ramsey sentence is "whatever it is" that satisfies the relations specified by the sentence. This 'indirect' reference is achieved by 'purely logical terms (variables, quantifiers, etc.) plus terms whose *direct* referents are items of acquaintance [i.e. observables]' (Maxwell 1970a, p. 16). One does not know what the natures of the unobservable entities are, but one can assert that they exist and stand in certain relations. 'This . . . may be taken as an explication of the claim of Russell and others that our knowledge of the theoretical is limited to its purely structural characteristics and that we are ignorant concerning its intrinsic nature' (Maxwell 1970b, p. 188).

If one adopts Maxwell's approach purely as a means of implementing Russell's, however, resulting in a knowledge of higher-order ('purely structural') properties as opposed to first-order ('intrinsic nature') properties and relations, it too is susceptible to the Newman objection. Some dispute this charge. Michael Redhead (2001a), for example, claims that the objection is ineffective, because it correctly applies only to positions that deny the reality of first-order properties and relations. Russell and Maxwell do not do this, and can admit that such properties and relations are instantiated while maintaining that one does not have any qualitative knowledge of them. Merely invoking the existence of first-order properties and relations, however, does not dissolve Newman's worry. As Newman (1928, p. 140) himself points out, it is only by knowing these properties and relations, as opposed to knowing merely of their existence and their higher-order properties, that one can distinguish between substantive as opposed to trivial instantiations of Russell's structures.

Worrall and Elie Zahar (2001) argue that the Newman objection is effective only because Russell speaks as though our knowledge of reality is *purely* structural, but their structuralism applies only to our knowledge of unobservables. If a theoretical description includes observable terms, they say, the corresponding Ramsey sentence is not trivially satisfied. This response on behalf of epistemic SR, however, is little comfort, for now the only condition on the satisfaction of a Ramsey sentence, beyond the constraint of cardinality, is that it be empirically adequate. In other words, it is enough that its observable consequences be true. Furthermore, Jane English (1973) has shown that any two Ramsey sentences that are observationally equivalent are consistent with each other. Since Ramsey sentences with the same observable consequences do not conflict, theoretical equivalence here amounts to nothing more than an equivalence of observable consequences. So much for this as a proposal for scientific realism! Recall that realism aspires not merely to a knowledge of observables, but to a knowledge of unobservables as well. This goal is not served by an account on which theories are true merely in virtue of being empirically adequate, and neither is it served by a view on which the content of a theory amounts to its observable consequences.

I will not discuss the Newman objection further here. The debate surrounding it is interesting in its own right, but so far as the realist is concerned, it is very much a red herring. The kind of knowledge epistemic SR claims regarding unobservables falls well short of anything resembling scientific realism. By hitching its wagon exclusively to a knowledge of higher-order, formal properties, epistemic SR no longer represents a proposal for realism. Luckily for the realist, however, this infatuation with structure as a higher-order property is entirely avoidable. In the same paper in which Newman demonstrates that Russell's concept of structure applied to the epistemic context yields trivial knowledge at best, he notes that his objection does not arise where one has some qualitative knowledge of the actual (first-order) relations between things. Is there an account of structure that incorporates this sort of knowledge? I believe there is. The applicability of the Newman objection turns on the question of what sort of structures one thinks one can know. Despite his fidelity to Russell, Maxwell (1970a, p. 17) himself hints at an appropriate direction for the realist, here: 'Causal connection must be counted among these structural properties, for it is by virtue of them that the unobservables interact with one another and with observables and, thus, that Ramsey sentences have observable consequences.' In suggesting the importance of causal connections, Maxwell was on to something of great importance to realism.

2.3 SEMIREALISM (OR: HOW TO BE A SOPHISTICATED REALIST)

Let us take Russell's definition of structural identity as furnishing a necessary, but not a sufficient condition of the notion of structure required by the realist. The sort of structures realists are after, I suggest, are not higher-order properties – properties *of* relations between first-order properties of entities (like the property of total ordering) or properties *of* relations between the entities themselves. Instead, let us identify structure *with* relations between first-order properties. Recall the ordinary, intuitive notion of structure with which we began. To give the structure of something is to enumerate its parts and describe the relations between them. Consider, for example, the structures of things like tables, causal processes, and societies. The concept of structure required by the realist is one that is tied to specific kinds of relata and their characteristic relations. As will become clear in what follows, the relata I have in mind here are generally, in the first instance, quantitative and determinate properties (such as masses, charges, volumes) of particulars (such as objects, events, processes). How kinds of particulars are then constituted from kinds of properties is a further question, a full answer to which must wait until a discussion of properties and natural kinds in Part II. In the meantime, however, let me attempt to clarify this notion of structure at the heart of scientific realism.

First, it may help to distinguish the idea of structures tied to kinds of relations and relata from the more generic concept Russell employs. Redhead (2001a, 2001b) offers some useful terminology here by distinguishing between *abstract* and *concrete* structures. Abstract structures are precisely what Russell describes: higher-order, formal properties of relations. But one and the same abstract structure may be instantiated by many different concrete structures, as we saw in the example of the cakes and the spoons.[4] A concrete structure consists in a relation between first-order properties of things in the world. Of course, when the first-order properties of things are related in a certain way, the particulars having these properties are often related in a correlated manner. If the mass of one cake (a determinate property) is greater than the mass of another, the former cake (a particular object) is heavier than the latter. This straightforward sort of correlation does not apply in all cases, but it does apply quite generally to many of the quantitative relations of interest to the sciences, and I will take it for granted in what follows.

Thus, when one says that blueprints display the structure of a house, or that the structure of DNA is revealed by Watson and Crick's demonstration model of the double helix, these are instances of an identity of abstract structure. In order to construct many of these sorts of representations, it is necessary only that the representation instantiate the same abstract structure as that which is represented, or something close, since representations are rarely perfect. Two concrete structures having the same higher-order, formal properties, but concerned with different sorts of

---

[4] The language of instantiation here may suggest a commitment to universals, but this is merely expedient. One might speak of classes of concrete structures instead. I will continue to talk of properties *simpliciter*, but everything said in this connection can be understood in terms of transcendent universals, immanent universals, tropes, or resemblance nominalism.

relata and relations, however, are *not* one and the same concrete structure. The molecules composing strands of DNA and the materials composing models of such strands are different sorts of relata, and thus comprise different concrete structures. Realists have traditionally aspired to a knowledge of the concrete. An identity of concrete structure requires that the elements of the sets compared, α and β, as well as their respective relations, R and S, be of the same kind. Theories generally describe relations between relata, first-order properties, in terms of mathematical equations in which the variables name kinds of properties. Advocates of epistemic SR think that only by restricting knowledge to abstract structures can realists respond to worries such as PI. But realists, I will suggest, can be more ambitious than this, with no consequent impairment to their response to antirealist scepticism – on the contrary. When I speak of structure henceforth, unless otherwise indicated, it is concrete structure I intend.

Concrete structures are relations between first-order properties of things, so to know them is to know something qualitative about relations, not merely their higher-order properties. For this reason, a knowledge of concrete structures is immune to the Newman objection. We will need a name to differentiate this understanding of realism from the epistemic structuralism we have been considering. Like ER and epistemic SR, a view that aspires to a knowledge of concrete structures may be regarded as a form of selective scepticism, for as we shall see, much of what is described in theories exceeds the concrete structures to which such a realist should commit. The rest of this chapter is devoted to an outline of this account of realist selective scepticism. Let us call it 'semirealism'.

It is time now to take up Maxwell's hint that causal connection is a crucial aspect of structure. The first-order properties whose relations comprise concrete structures are what I will call *causal* properties. They confer dispositions for relations, and thus dispositions for behaviour on the particulars that have them. Why and how do particulars interact? It is in virtue of the fact that they have certain properties that they behave in the ways they do. Properties such as masses, charges, accelerations, volumes, and temperatures, all confer on the objects that have them certain abilities or capacities. These capacities are dispositions to behave in certain ways when in the presence or absence of other particulars and their properties. The property of mass confers, *inter alia*, the disposition of a body to be accelerated under applied forces. The property of a volume on the part of a gas confers, *inter alia*, the disposition to become more highly pressurized under applied heat, and so on. It is the ways in which these dispositions are linked to one another – that is, the ways in which particulars with various properties are disposed to act in consort with others – that produce causal activity. Causation, ultimately, has to do with relations determined by dispositions, conferred by causal properties.

Talk of dispositions is second nature to some and worrying to others. A more thorough consideration of dispositions will occur in Chapter 5, but for now, let it suffice to say that no one need be concerned by the invocation of dispositional ascriptions at this stage. Realists about dispositions, or those who believe there are causal powers in nature, may skip to the next paragraph. Empiricists, however, who in the spirit of Hume are traditionally wary of what they take to be the mysterious connotations of powers, may take some comfort here. There are well-rehearsed ways in which disposition-talk may be elaborated, and though some happily take it at face value, empiricists may opt for a deflationary analysis of dispositional language in terms of conditionals if they wish. (To say that a substance is soluble is just to say that it dissolves if placed in an appropriate solvent, and so on; Mumford 1998, ch. 3, summarizes the difficulties faced by such approaches.) For present purposes I will merely say that properties confer dispositions to enter into relations. The ambiguity of 'confer' here is intended to signal neutrality on the precise details, ontological and otherwise, concerning dispositional ascriptions. The present contention is simply that properties are responsible for the behaviours of things. Particulars behave as they do because they have causal properties.

One important feature of semirealism is that the central thesis of epistemic SR, that one can have knowledge of structures without knowledge of the intrinsic natures of things, cannot be maintained. Epistemic SR commits to abstract structures and not the intrinsic, but semirealism rejects this prescription, for a knowledge of concrete structures *contains* a knowledge of intrinsic natures. Concrete structures are identified with specific relations between first-order properties of particulars, and first-order properties are what make up the natures of things. So on this view, to say that two sets have the same structure is *ipso facto* to say something about the intrinsic natures of their members. Furthermore, concrete structures arise as a consequence of the dispositions conferred by these first-order properties. Natures are thus intimately connected to the relations into which properties and particulars enter. Speaking rather loosely, one might say that while causal properties are intrinsic, they also have a "relational" quality. They are "relational" in that they confer dispositions, and dispositions determine the sorts of relations properties and particulars can enter into. Knowledge of these relations thus gives the realist insight into the intrinsic natures of things.[5]

According to semirealism, then, structure and nature come together, but this should not be taken to suggest that they are properly conflated. Stathis Psillos (1995; 1999, ch. 7), for example, holds that there is no principled distinction between the structure and the nature of an entity or process, and that this is one reason to reject epistemic SR. I am sympathetic to his conclusion, but I suspect this is not the way to argue for it, for there is a genuine distinction between the natures of particulars and structural relations. Some structural aspects of a thing may constitute part of its nature, just as the specific bonding relations between hydrogen and oxygen atoms constitute part of the nature of water. But the relations of which particulars are capable exceed such descriptions of intrinsic natures. A particular's nature comprises its first-order properties, and natures are possessed whether or not particulars are, at any given moment, manifesting all of the relations of which they are capable. Since structures are identified with relations, there are no structures to speak of unless these relations obtain. Thus, particulars always have natures (first-order properties), but whether or how they can be described structurally will depend on what relations they happen to be manifesting at any given time, and these may vary, depending on the circumstances. Nevertheless, given that causal properties are understood in terms of dispositions for relations, structural knowledge does contain a knowledge of properties, and thus natures. Structures are, metaphorically speaking, "encoded" in the natures of particulars, because first-order properties confer dispositions for specific relations – those one recognizes as structures.

Some may worry, though, about the move to concrete structures. After all, was it not the point of SR to do away with intrinsic natures, to speak of relations between things in the world, and not the things themselves? By denying that realists should attempt to separate a knowledge of relations and relata in this way, one might argue that I am no longer describing a form of structuralism. If semirealism permits knowledge not only of structures but of natures and causal relations, one might think this a *reductio* of the very idea of structuralism. On such a view, suggests

---

[5] Interestingly, Maxwell (1970a, pp. 33–4, n. 19), mentions dispositions *en passant*. He says that if one were to redefine higher-order property terms by means of 'a viable causal redefinition', one could predicate them of entities. For example, one might redefine 'red' so that it refers, not to a property of visual experience, but rather to a disposition on the part of objects that appear red to us. But he concludes that such a disposition, though a structural property, would not be a first-order property of the objects in question. This is precisely what semirealism denies.

David Papineau (1996, p. 12), 'restriction of belief to structural claims is in fact no restriction at all'. The worry is that defining structures in terms of relations between first-order properties weakens the structuralist component of semirealism to the extent that it collapses back into a more comprehensive scientific realism, as opposed to a form of selective scepticism.

This charge, I think, has things back to front. It does not appreciate the evolution of scientific realism – what it was and what it has become. "Standard" realism is not what it once was; it has been refined in response to antirealist scepticism. The extent to which sophisticated realists today are able to respond to arguments such as PI, I believe, is proportional to the extent to which they have moved, whether they realize it or not, towards a form of structuralism. Richard Boyd (1981, pp. 613–14) characterizes realism by saying that 'typically, and over time, the operation of the scientific method results in the adoption of theories which provide increasingly accurate accounts of the causal structure of the world.' Psillos (1999, p. 155) describes scientific knowledge this way:

When scientists talk about the nature of an entity, what they normally do – apart from positing a causal agent – is to ascribe to this entity a grouping of basic properties and relations. They then describe its law-like behaviour by means of a set of equations. In other words, they endow this causal agent with a certain causal structure, and they talk about the way in which this entity is structured.

Various sophisticated accounts of realism, I contend, gesture in significant ways towards structuralism. But semirealism extends this movement beyond these and more full-blown accounts generally in two important respects. The first concerns the fact that although some contemporary realists have placed ever greater emphasis on causal structures, they generally give little or no consideration to what causal structures are, precisely. Unless careful attention is given to this question, realists cannot hope to understand where to draw the line between aspects of theories they have good reason to endorse and those they do not, and this leads to the second way in which semirealism reaches further. Merely invoking causal structures is not enough, for not all structures described by theories are worthy of realist commitment. A selective sceptic must know how to be selective, and elaborating this advice is the task to which I will turn now.

Given a knowledge of concrete structures, it is no miracle, claims the realist, that good scientific theories are empirically successful, for they describe the structures of reality. In order for semirealism to count as a form of selective scepticism, however, it must offer a principled means of distinguishing parts of theories that are likely to be retained as the sciences move on from those that are apt for replacement. And in order for it to count as an *effective* form of selective scepticism, it must offer some reason to think that this demarcation constitutes a compelling response to PI. Let us see whether semirealism is up to these tasks.

2.4 OPTIMISTIC AND PESSIMISTIC INDUCTIONS ON PAST SCIENCE

On further reflection, it is surprising that anyone thought that epistemic SR might rescue realism from worries about PI. Indeed, the same will go for ontic SR; as we shall see in Chapter 3, the motivation for this position comes more from modern physics than from PI, but in any case it is likewise helpless to respond. The reasons for this failure are not exhausted by the fact that their conceptions of structure are problematic. Imagine that epistemic SR was somehow free of the difficulties I have outlined. Even then it would not offer the realist a helpful response. Poincaré and Worrall following him gesture towards knowledge contained in the mathematical equations of theories (subject to normal realist caveats). While this may seem promising in the case of the specific equations cited by Poincaré, developed by Fresnel, it is not at all promising in other cases of equations that were then a common part of theorizing about light. Worrall is quick to note that Fresnel's equations can be derived from Maxwell's theory of electromagnetism, but many scientists at the time of Fresnel were hard at work on models of the ether, in which various relations between its imagined properties were described mathematically. Needless to say, *these* equations are not retained in the later theory of electromagnetism. Unfortunately for SR, scientists and theories generally describe many structures, but only some of these are likely to be retained. The realist needs a way to determine which these are, but she will not learn how to do so from either epistemic or ontic SR.

Other realists have attempted to differentiate aspects of theories most likely to be retained by appealing to the idea that only some aspects are required to 'do the work' of scientific prediction, retrodiction, and explanation. Other parts of theories, they claim, are not essential for these tasks. Kitcher (1993, pp. 140–9), for example, distinguishes between what he calls the 'presuppositional posits' or 'idle' parts of theories, and 'working posits', the parts that are required to generate predictions and explanations. Psillos (1999, p. 108) claims that 'it is enough to show that the success of past theories did not depend on what we now believe to be fundamentally flawed theoretical claims ... it is enough to show that the theoretical laws and mechanisms which generated the successes of past theories have been retained in our current scientific image'. He considers various incarnations of the caloric theory of heat and optical ether theories, arguing that the conceptions both of caloric as a fluid and of the ether as an elastic solid, now discarded, were merely heuristic aids as opposed to essential parts of past theories. The essential bits, he maintains, are retained in current theory.

As it stands, however, this sort of advice is not specific enough to be especially helpful. It is not clear what criteria the realist should use, in general, to separate the wheat from the chaff. Psillos hopes to take direction from the relevant scientists themselves in order to aid the realist here, but this is a dubious methodology. For not only does it require that one know the ontological commitments of past scientists, which is often not at all transparent from their writings, but scientists themselves commonly have widely diverging views on such matters. Lacking more helpful direction on how to proceed, Kitcher and Psillos leave their approach open to the charge of rationalization *post hoc*. It is too easy, one might claim, to "identify" those aspects of past theories that did the "real work" after the fact. Looking back from the perspective of the present, one is bound to think of parts of past theories that have been retained as those that were indispensable, given that other aspects have been dispensed with in current theory. From the perspective of the present, so the charge goes, it is only natural that one identify retained elements as required elements.

I will return to the problem of *post hoc* rationalization shortly, but first let me attempt to improve on the weaknesses of this last strategy on behalf of the realist. Case studies are certainly valuable as confirmatory illustrations, but the realist requires a general account of what it means to 'do the work' one asks of theories. Like ER and SR, the broad appeal to the working parts of theories, on its own, is insufficient to yield a compelling account of selective scepticism. Semirealism, on the other hand, furnishes an *a priori* reason for thinking that certain structures will be retained. Recall that semirealism takes inspiration not only from SR, but also from ER. It is time now to recall the epistemic lessons of ER from the beginning of this chapter.

Earlier I mentioned various realist responses to PI, but committed to investigating the policy of selective scepticism in particular. On this latter approach it is granted that there is a significant amount of discontinuity in scientific theorizing over time, which seems undeniable. What there is not, however, is much in the way of radical discontinuity in what is *properly believed*. The trick is to separate aspects of theories most worthy of belief from those for which one has less warrant. Once this distinction is made, realists can admit a pessimistic induction on the history of past science *simpliciter*, while simultaneously asserting an optimistic induction on the parts of theories to which they commit, to the extent that these parts tend to survive over time. The structures to which realists should commit, echoing the most persuasive insight of ER, involve properties and relations that are essential to describing our causal connections to the world. Realists should commit to concrete structures that are detected and described in scientific theories. It is important to keep in mind, here, one of the concluding morals of the preceding discussion of ER: realists must be realistic. There is no question of a strict delineation of structures into those one knows and those one does not. A realist's degree of belief should reflect one's degree of causal contact, with mastery and manipulation at one end of the spectrum, and mere detection and weaker speculation at the other. I will make no further mention of this qualification henceforth, but take it as understood.

To facilitate this discussion, let me introduce a distinction between the attribution of *detection* properties, and *auxiliary* properties. Detection properties are causal properties one has managed to detect; they are causally linked to the regular behaviours of our detectors. Auxiliary properties are any other putative properties attributed to particulars by theories. This is an epistemic distinction. Detection properties are the causal properties one knows, or in other words, the properties in whose existence one most reasonably believes on the basis of our causal contact with the world. The ontological status of auxiliary properties is unknown – they may be causal properties, or fictions. An auxiliary property is one attributed by a theory, but regarding which one has insufficient grounds, on the basis of our detections, to determine its status. Whether the attribution of a property qualifies as detection or auxiliary property attribution will depend on the state of scientific inquiry at the time. As the sciences move on, some auxiliary properties are retained as auxiliary, some are converted into detection properties, and others are simply discarded.

Causal, detection, and auxiliary properties are related to each other in several ways, as shown in Figure 2.2. All detection properties are causal properties. To detect is to establish a causal link with the particular under investigation. The attribution of auxiliary properties, however, is

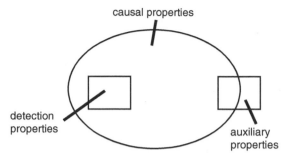

Figure 2.2. Property distinctions underlying semirealism

noncommittal with respect to their ontological status. Further investigation may allow us to detect them, thus converting them into detection properties, or may rule them out altogether. Causal properties themselves are not exhausted, of course, by those attributed to particulars by theories. If it turns out that detection content is generally retained as theories change over time, and what is left behind is generally auxiliary, the realist would have a systematic basis for an account of theoretical knowledge, past and present. A realist could then commit to relations of detection properties, and remain agnostic or sceptical about auxiliary properties. All that remains to complete this sketch of semirealism, then, is to explain how the realist can identify concrete structures having this epistemic warrant, and to explain why precisely these structures are likely to be retained as theories change. Let us turn to these matters now.

The realist requires a practical means of demarcating detection properties (and the structures associated with them) from auxiliary properties. Here is a suggestion. Detection properties are connected via causal processes to our instruments and other means of detection. One generally describes these processes in terms of mathematical equations that are or can be interpreted as describing the relations of properties. As I will attempt to show, one can thus identify detection properties as those that are required to give a *minimal interpretation* of these sorts of equations. Anything that exceeds a minimal interpretation, such as interpretations of equations that are wholly unconnected or only indirectly connected to practices of detection, goes beyond what is minimally required to do the work of science: to make predictions, retrodictions, and so on. The excess is auxiliary.

Recall Poincaré's and Worrall's example of the transition in theories of light from Fresnel to Maxwell. Here is the system of equations Fresnel developed to describe the circumstances represented in Figure 2.1:

$$R/I = \tan(i - r)/\tan(i + r)$$
$$R'/I' = \sin(i - r)/\sin(i + r)$$
$$X/I = (2\sin r \cdot \cos i)/(\sin(i + r)\cos(i - r))$$
$$X'/I' = 2\sin r \cdot \cos i/\sin(i + r)$$

Light can be analyzed as a wave-type disturbance. Consider an ordinary beam of unpolarized light. (Speaking somewhat loosely, 'unpolarized' simply means that the "vibrations" of the wave, which are perpendicular to its direction of propagation, are in no one uniform direction, but in many directions.) The polarization of such a beam can be resolved into two component planes, at right-angles to each other. One of these is called the plane of incidence, and contains the incident, reflected, and refracted beams (the plane of the page, in Figure 2.1). The other component is polarized in a plane at right-angles to the incident plane. I^2, R^2, and X^2 represent the intensities of the incident, reflected, and refracted components respectively, polarized in the plane of incidence. I'^2, R'^2, and X'^2 represent the intensities of the components polarized at right-angles to the incident plane. i and r represent the angles made by the incident and refracted beams with a normal to the plane of reflection.

The existence of certain properties is minimally required to give a realist interpretation of these equations, viz., intensities, and directions of propagation. These are first-order, intrinsic properties of light, but what about the ether, or the electromagnetic field? In the very limited context of these specific equations, ethers and fields are auxiliary posits. Our theories incorporate such entities as important heuristic devices; they help to fill out one's conceptual pictures of the phenomena. In keeping with the idea of selective scepticism, however, the realist should be wary here. The advice semirealism gives is straightforward: believe in the relations of detection properties, as given in the minimal interpretation, and treat anything that exceeds these structures with caution. Furthermore, recall that these properties are dispositions. When light is subjected to certain forms of detection, certain concrete structures are manifested. Present-day theory describes light in terms of quanta called photons, viewed as excitation states of a particular sort of field as opposed to anything resembling an ordinary or classical wave. Nevertheless, light is no less disposed to give rise to detections of wave-type behaviour today than it was 200 years ago.

The question of how the realist is to arrive at a minimal interpretation is crucial here, and requires further consideration which I will undertake in the next section. First, however, let us consider why descriptions of detection properties and their relations, identified by giving such interpretations of mathematical equations connected to detection, are likely to be retained in some form in later theory. The answer is that one simply cannot do without them. These equations (or ones that approximate them) are *required* to describe the regular behaviours of our detectors. If realists interpret this mathematical formalism in terms of concrete structures, then again, recalling the insight of ER, it is these structures to which they have the best epistemic access, for these structures are causally connected to our means of detection. It is thus no surprise to realists that descriptions of the structures to which they have the best epistemic access should remain relatively stable as theories are modified and improved over time.

Semirealism thus empowers the realist to make surprisingly strong claims. In most cases one *must* retain specific structures involving detection properties, or something very much like them, if one is to retain the ability to make decent predictions. Consider the theory of electromagnetism. No matter what form the descendants of this theory take, one would presumably lose the result that the speed of electromagnetic radiation is c if one did not retain something like Maxwell's equations as a component of these descendants.[6] Of course, current theories do not retain all of the structures described by their predecessors. But not all structures are causally connected to our practices of detection. The realist should expect to retain only those structures required to give a minimal interpretation of the mathematical equations used to describe well-established practices of detection, intervention, manipulation, and so on. The semirealist thus anticipates both optimistic and pessimistic inductions on the history of science. And though many contemporary theories will not survive in their current form, one should not be too disparaging of their auxiliary content. Given the important heuristic role played by auxiliary properties, one should expect and commend their presence in theories.

In a classic paper on inter-theory relations, Heinz Post (1971) argues that heuristic principles typical of scientific investigation promote theories that 'conserve' the successful parts of their predecessors. He thus defines his

---

[6] This was suggested by Steven French in correspondence, which is especially kind given my discussion of his position, ontic SR, in Chapter 3. See Saatsi 2005 for an excellent study of retention focusing on the Fresnel-Maxwell case study.

'General Correspondence Principle': new theories generally account for the successes of the ones they supersede by 'degenerating' into the older theories under conditions in which the older ones are well confirmed. Post's examples come primarily from the history of physics, but realists commonly maintain that this sort of correspondence is typical of the chemical and biological sciences also (French and Kamminga 1993 contains several such studies). Indeed, any version of realism will need to give an account of this sort, because the situation described by Poincaré with regard to Fresnel and Maxwell is misleadingly atypical. In this particular instance, Fresnel's equations are neatly and strictly preserved, but many cases of scientific succession are not like this. It is hardly rare for the structures of earlier theories merely to approximate those of their successors, or to describe only special cases of later theories. Here too, realists must be realistic. So long as relations of correspondence and approximation obtain, however, and there is significant evidence that this is the norm, the realist story is there for the telling.

Many antirealists are keen to dispute this. In one of a number of similar examples, Laudan (1981, pp. 39–43) claims there can be no continuity of the sort Post describes between classical and relativistic mechanics, for although some laws of the former are limiting cases of laws of the latter, others cannot be, because they invoke an entity that is simply not countenanced by the later theory: the ether! But *of course* there are parts of theories that later theories do not retain – that, in addition to incorporating new content, is the very point of scientific change. This is not at all surprising or worrying to the sophisticated realist. The ether is part of the auxiliary content of earlier theories. It is not required to give a minimal interpretation of equations describing relations between detection properties. Semirealism has nothing invested in the ether.

The notion of a minimal interpretation of equations, however, on which the identification of detection properties and associated structures depends, bears a great epistemic weight here. In many cases, determining the minimal interpretation may seem a considerable challenge. Furthermore, if structures are to be retained in some form from one theory to the next, it is important that the reference of terms for the relevant properties and relations is maintained across theories. Since theoretical descriptions of these structures are refined over time, this will require a theory of reference with a causal component. These issues raise two concerns. Firstly, one might argue that the task of giving a minimal interpretation is impossible in anything other than retrospect, where the inevitable use of hindsight raises suspicions about the identification of the relevant structures.

Secondly, one might hold that the appeal to some version of the causal theory of reference makes it impossible that the unobservable terms in most theories could fail to refer. These worries are linked, and threaten to turn semirealism into an empty position: uninformative and trivially satisfied. Let us consider these charges in more detail.

2.5 THE MINIMAL INTERPRETATION OF STRUCTURE

Hindsight is 20/20. As mentioned earlier, one might claim that it is all too easy to "identify" the parts of past theories responsible for their successes in retrospect. Looking back, one then "discovers" that these parts are the ones to which realists should commit. The accusation here is of rationalization *post hoc*. From the perspective of the present, it is no wonder that realists identify aspects of past theories that have been retained as those that do "the real work" of the sciences. If this is correct, the realist's commitment to these aspects is suspect. It would seem that he or she believes in retained elements because they are retained, not because they are more likely to be true or approximately true. And if this were the case, it would be impossible to know at any given time what new structures posited by current theories should be believed – such determinations could be made only in retrospect.

To defend themselves against these doubts, realists must do more than merely identify belief-worthy parts of theories with parts that are preserved. That is, they must do more than merely suggest a correlation. They must explain why the correlation obtains. Semirealism does this for the realist because it offers a formula for identifying the belief-worthy parts of theories, and this formula is recommended on the basis of its epistemic value. When realists determine which relations are minimally required to interpret the mathematical formalism of a theory in the context of detection, and thus identify concrete structures, their belief in them stems from the fact that these are the structures that cannot be denied if one accepts that the theory is reasonably successful in describing parts of the world and their relations to our detectors. These are the aspects of theories for which one has the greatest epistemic warrant, because these are the aspects that cannot be done without in making predictions, retrodictions, and so on. Antirealists will have different views, here, regarding what constitutes an appropriate minimal interpretation, and consequently what is warranted. But my present concern is with realism. When they identify detection properties, whether in new theories or in old, realists have *a priori* reason to believe that descriptions of them will be retained in some form.

No doubt, this would-be recipe for success – the minimal interpretation – will in many cases prove more circumspect than an old-fashioned realist might like. Theories often describe quite general causal frameworks, and the portion of a theory to which the realist should commit may be embedded in a larger framework of this sort. Consider once again the Fresnel-Maxwell case study. Though Fresnel had a particular view of a general causal framework involving the behaviour of light, not all of this understanding is required to give a minimal interpretation of his equations. Fresnel thought that light is a disturbance in the ether, an elastic solid medium, but the semirealist is wont to say: 'Wait, slow down, compose yourself, *look at the equations.*' Lacking an appropriate notion of structure, this is unhelpful counsel at best, but given the concept of concrete structure, the realist can make something of this advice.

So let us examine the equations and consider the properties one finds described there. Minimally, the variables simply represent amplitudes (intensities) and angles (directions of propagation). But are these not, as Fresnel believed, intensities and directions of propagation in the ether? The semirealist is unmoved by this appeal to the greater causal framework. The variables name properties, and these properties are to be understood simply in terms of dispositions to enter into the very relations of properties described, in summary form, by these equations. To suppose that a direction of propagation is *furthermore* a direction in the ether is to go beyond what is minimally required to give an interpretation of this particular set of equations. To suppose that accelerations are *furthermore* accelerations with respect to absolute space, as Newton did in contemplating his second law of motion, is to go beyond what is minimally required to interpret the relations between properties described by the expression '$F = ma$'. For the realist, here, less is more. The semirealist thus commits to relations of intensities, directions, masses, accelerations, and so on, and remains agnostic or sceptical about any further embellishments.

This prescription may be difficult to follow for some, because it asks the realist to refrain from commitments to parts of theories that do play explanatory roles, and realists have a weakness for explanations. For Fresnel and other ether theorists, the causal story told by their equations is *ipso facto* part of a causal story involving the ether. The semirealist asks for a suspension of belief: in this case one is asked to separate various aspects of the overarching causal story of Fresnel's theory, and to believe only those one cannot do without in giving a minimal interpretation of his equations. This involves separating aspects of theories that for psychological, professional, theological, or other reasons, the scientists who develop and use these theories may have difficulty disentangling. But it is not impossible. Indeed, we know this to be true, since Fresnel's equations were ultimately accepted as part of Maxwell's theory in the context of a non-ethereal physics. The recipe of the minimal interpretation is austere, and straightforward: commit only to structures with which one has forged some significant causal contact, and understand the natures of detection properties in terms of dispositions for relations to other properties.

On this account of realism, however, reference to the relevant structures cannot be fixed by description alone. As noted earlier in the context of the Correspondence Principle, theoretical descriptions are generally refined as theories improve, yet realists hope to refer to the same properties and relations throughout, over time. It seems they require something like a causal theory of reference, where links between language and world are created in baptismal events, in which baptizers are causally connected to the items named, and future uses of these terms are parasitic on these events. Mixed theories of reference such as causal-descriptive theories would also serve here. According to causal-descriptive theories, reference is fixed by descriptions of some core properties of the thing referred to, which are responsible for the causal processes in virtue of which it is baptized. I will not consider the issue of reference in any detail here, but it is important to draw attention to it, because it is a commonly heard complaint that causal theories trivialize reference. If one can ensure that one refers by assigning the referents of terms vague enough causal roles or properties, this would seem to make error impossible, and no realist should claim that one's commitment to structures is infallible.

So long as realists are alert to the threat of triviality, however, semirealism supplies conceptual tools with which to overcome it. Responding to allegations that the central terms of many past theories are non-referring, Clyde Hardin and Alexander Rosenberg (1982) suggest that one possibility for the realist is to assert that many such terms *can* be construed as referring, on the basis of the causal roles attributed to their referents by the theories in which they occur. This is a variation on Putnam's (1978, pp. 22–5) more general notion of the principle of the 'benefit of the doubt'. One may construe Mendel's use of the term 'gene' as referring, they say, even though nothing called a 'gene' in contemporary genetic theory answers to the description Mendel gave, because in current theory configurations of DNA and their polypeptide products perform the same causal role as that attributed to genes by Mendel.

As a general strategy for realism, however, this is too permissive. A realist must know where to draw the line between reasonable applications of the principle of the benefit of the doubt and applications that go too far, attributing approximate truth to arcane theories merely because their ontologies were posited to account causally for some of the same phenomena that interest us today. Semirealism again provides some useful advice here. It is reasonable to give the benefit of the doubt in cases where not just general causal roles are retained, but where quite specific dispositions for relations conferred by particular detection properties are preserved. On this view, it would be unreasonable to apply the principle in such a way as to identify (with one another) the putative referents of significantly different systems of properties.

Perhaps an example will help to clarify this advice. Consider an infamous case of theory change for the realist: the eighteenth-century transition from Georg Stahl's theory of phlogiston, defended by Joseph Priestley, to Antoine Lavoisier's theory of oxygen. Phlogiston theory accounts for combustion, calcination (rusting), and respiration in terms of the removal of a colourless, odourless, tasteless substance, phlogiston, from particulars undergoing these processes. Oxygen theory accounts for the same phenomena in terms of the absorption of oxygen. Present-day chemistry denies the existence of phlogiston. Yet 'dephlogisticated air' might refer, not to air that is lacking in phlogiston, conceived as a real substance, but rather to oxygen. After all, one might say, it is important to distinguish between truth and reference. Successful reference does not require that the referring expression is true of its referent. One often refers to things successfully with incorrect or merely partially correct descriptions. And so, one might conclude, the realist should be happy to say that the greater the extent to which air is low in phlogiston, the greater the extent to which it is rich in oxygen, and *vice versa*, with no unease about the reference of these expressions.

This is precisely the sort of conclusion that dismays critics of the realist appeal to causal theories of reference. It trivializes reference, they claim, to say that Priestley was talking about oxygen all along. There is a sense, of course, in which Priestley *was* talking about oxygen (assuming that his experiments involved *inter alia* the presence and absence of oxygen as opposed to phlogiston), but realism appears ridiculous if one says that assertions regarding phlogiston and oxygen are mere linguistic or notational variants of one another. As far as semirealism is concerned, however, there is no question of identifying Priestley's descriptions with Lavoisier's. For although some of the causal roles described by their theories for dephlogisticated air and oxygen are the same, it is plainly not the case that the putative causal properties of dephlogisticated air are co-extensive with those of oxygen.

How does one know this, and how might realists have known this at the time of Priestley and Lavoisier? Oxygen has a fixed chemical composition, but dephlogisticated air, *ex hypothesi*, does not. Different combinations of gases may lack phlogiston, and different combinations of gases have radically different dispositions. The same cannot be said of oxygen. Clearly, then, one is dealing here with very different sets of putative detection properties. As a consequence, as suggested by semirealism, the realist should not accept the claim that 'dephlogisticated air' and 'oxygen' refer to the same thing. The clarity of hindsight may assist, of course, in determining that these expressions should not be thought of as co-extensive. In such cases, however, retrospection is no cause for concern, since the very point of the exercise is to consider cases of theoretical change – something that can be done only in retrospect. Semirealism thus provides an antidote to the threat of triviality.

I have outlined a form of realism, taking inspiration from the best insights of ER and epistemic SR, and learning, I hope, from their missteps. This I believe is the natural course of the evolution of realism, and to the extent that sophisticated realists are able to respond to instances of antirealist scepticism such as PI, it is because they have moved implicitly in this direction. The central idea of semirealism is that one can apply the epistemic lessons of ER, regarding the significance of our causal interactions with the world, in such a way as to understand what is right about the appeal, by SR, to the notion of structure. A knowledge of abstract structures not only is problematic in the scientific context, but also gives too weak a purchase on reality to constitute much of a realism. The evidence of mathematical continuity across theories commonly cited by structuralists, however, supports more than this. Realists can know first-order properties and relations by minimally interpreting the mathematical equations one uses to describe scientific investigations into the nature of the unobservable.

On this basis, it is arguable that one can also know conservation laws and other principles such as, for example, the Pauli exclusion principle, which states that no two fermions (a category of subatomic particles) in a closed system, like two electrons in an atom, can have precisely the same state (the set of properties described by their quantum numbers). These are not concrete structures *per se*, but they may be ways of summarizing aspects of structural relations. These and other questions remain to be settled by realists on a case-by-case basis, and there is, no doubt, room for debate and disagreement. In all such cases, however, debates are grounded in the fact that descriptions of concrete structures underdetermine not the detection properties of particulars and their relations, but simply the auxiliary content of theories, and this demarcates plenty of theory about which realists may be cautious, agnostic, or sceptical, and gives ample reason to expect both optimistic and pessimistic inductions on the history of the sciences.

CHAPTER 3

Properties, particulars, and concrete structures

3.1 INVENTORY: WHAT REALISTS KNOW

Entity realism (ER) and epistemic structural realism (epistemic SR) are proposals for realist humility. They offer to distinguish parts of scientific theories that are good bets for knowledge from others that are less so, thus allowing the realist to come to grips with the fact that accepted theories change over time. Neither ER nor epistemic SR, however, is humble in quite the right way. Their prescriptions for how realists ought to be selective sceptics are problematic and ultimately, I believe, untenable. As steps in the evolution of realism, however, they are on the right track, and I have aimed to incorporate the best insights of both under the heading of 'semirealism'.

The lesson of ER concerns the epistemic basis of claims about unobservables. By emphasizing causation, ER captures the common and deeply held realist intuition that the greater the extent to which one seems able to interact with something – at best, manipulating it so as to bring about desired outcomes – the greater the warrant for one's belief in it. But ER attempts to separate a knowledge of entities from a knowledge of their relations, and this cannot be done. It also gives encouragement to awkward diagnoses of historical events. Imagine a review of the evidence considered by different physicists, over time, for thinking they had detected a negative charge. Is it correct to say they all believed in the same thing, the electron? Adopting ER, the answer appears to be 'yes', but this is too coarse a thing to say. In addition to a negative charge, these scientists associated many different properties with electrons. Enter semirealism, first and foremost a realism about well-detected properties. This refinement illuminates certain discriminations that are otherwise glossed over: they all believed in negative charge, and certain relations involving negative charge and particulars having it, but many of the other properties they associated with these particulars changed dramatically over the years as subatomic physics developed. And since on this view the realist understands properties in terms of dispositions for relations, there is no question of separating a knowledge of one from a knowledge of the other. A knowledge of entities and their relations is intimately connected here.

The lesson of epistemic SR concerns the fact that within domains of scientific theorizing, there appears to be a great deal of preservation of mathematical description over time. Such preservation, says SR, is surely not unrelated to the fact that theories in these domains continue to be successful. In suggesting that realists can thus know certain structures, however, the precise meaning of the term 'structure' becomes important, and, when clarified, its origins in Russell's epistemology prove ruinous, resulting in an inability to distinguish the structures realists should endorse from those they should not. Enter semirealism, where structure is understood in the everyday sense by enumerating the actual targets of scientific study (the properties of things) and describing their relations, as summarized in the mathematical formalism of successful theories. These relata are understood in a way that emphasizes their role in concrete structures – as dispositions to enter into relations of the types described. So when epistemic SR enjoins the realist to believe in relations but not relata, semirealists demur. They do not banish the relata, on pain of various problematic consequences, but rather understand them structurally, as dispositions to stand in the very mathematical relations SR deems important. And on this view, applying the lesson of ER and the minimal interpretation of structure, realists are in a position to separate what they believe from the auxiliary content of theories, thus allowing them to make sense of the history of the sciences.

The present chapter engages two main tasks. The first is to add some flesh to the bones of this sketch of semirealism, and the second is to defend it against a challenge offered by a final proposal for realist selective scepticism (mentioned earlier but deferred until now), ontic SR. In addressing the first of these tasks, I will argue in more detail that a knowledge of some of the properties described by theories and a knowledge of certain concrete structures presuppose and thus entail each other. Furthermore, lest anyone think that this sort of realism, with its talk of properties and structures, is too far removed from what one ordinarily thinks scientific knowledge is about, viz. various kinds of particulars investigated by the sciences, I will clarify how a knowledge of concrete structures is connected to talk about particulars: objects such as mitochondria and neutrinos, and processes and events such as cellular energy production and subatomic interactions.

Before tackling these subjects, one aspect of the semirealist approach requires clarification, for it has important consequences for the issue of how far the realist inventory extends. Thus far, I have spoken about a knowledge of properties and relations in terms of concrete structures, and have mentioned the apparently straightforward promise of extending this knowledge to generalizations summarizing features of the concrete structures we know, such as conservation laws and other principles. The means suggested for the acquisition of this knowledge involves the interpretation of mathematical equations. It is not immediately obvious, however, that all scientific subdisciplines, especially those outside of physics, characteristically employ mathematical formalism as their primary mode of description. One might argue that in molecular biology, for example, knowledge is not generally presented in the form of equations that can be interpreted as describing relations, and if this is the case it is unclear, perhaps, how semirealism is to be understood in such a context.

I think it is important to note here that the form in which scientific claims are expressed does not always reflect the form taken by claims that underwrite them, epistemically. And if asked to describe the knowledge a theory contains, one may well leave out details pertaining to the means by which that knowledge is ascertained. In many cases scientific subdisciplines do not present knowledge in the form of mathematical descriptions, when in fact such descriptions are precisely what underlie the knowledge claims presented. Consider the case of molecular biology. A theory may claim that under certain conditions, a particular kind of enzyme helps to catalyze the breakdown of a particular kind of biochemical compound – no mention of mathematical relations here! But realists are interested in how knowledge claims are justified, and the semirealist has a particular view of how realist justifications are achieved. Insofar as knowledge of a particular kind of enzyme is justified, it is based on detections of certain properties, and in order for a molecular biologist (let alone a realist) to be convinced that such things exist, there must be empirical tests in which these properties are detected. Detections are generally described in terms of quantitative relations. The same is true regarding knowledge of biochemical compounds, and processes such as catalysis. It is on the basis of relations between determinate properties that knowledge claims about biochemical objects and processes get their purchase. Mathematical relations are no less important here than they are in the case of knowing about electrons.

According to semirealism, realists should admit only properties and relations with which the sciences have forged some sufficiently significant causal contact, and in most cases this knowledge is justified by a knowledge of relations that can be described mathematically. Finely tuned participation in causal processes involving unobservables is usually an intricate business, and it would be very surprising indeed to learn that a molecular biologist was able to manipulate the levels of an enzyme in a cell, so as to bring about the desired rate of a catalytic process, without some knowledge of the quantitative relations between the relevant properties of the particulars involved. Earlier I said the semirealist takes the variables occurring in equations describing such well-detected phenomena as naming properties, and understands these properties simply in terms of dispositions to enter into the relations summarized by these equations. This is the minimal interpretation of structure. Any systematic dependence between properties can be described mathematically, whether in terms of simple equations like Fresnel's, relating the intensities and directions of incident, reflected, and refracted beams of light, or by different sorts of functions. In either case, the realist has a basis for a minimal interpretation.

In sketching the general framework of semirealism, I have laid out a catalogue of unobservable items for realists to believe in: properties, relations, structures, and as we shall see, a heterogeneous assortment of particulars. I have also said that it is first and foremost a realism about well-detected properties. Let us now consider these commitments and some important connections between them in more detail. More specifically, I will argue that a knowledge of properties of particulars brings with it, unavoidably, a knowledge of concrete structures, and concomitantly that a knowledge of structures brings with it, unavoidably, a knowledge of certain properties and thus of particulars.

3.2 MUTUALLY ENTAILED PARTICULARS AND STRUCTURES

Properties and relations between properties are different things. Knowledge of an unobservable property and knowledge of at least some of its characteristic relations, however, are not so easily separated. This point is already familiar from the previous discussion of ER, in which I argued that it is impossible to drive a wedge between knowledge of the existence of an entity and a knowledge of its relations. When it comes to unobservables, it is only *by means* of a knowledge of their relations that one is able to detect and manipulate them. In Chapter 5, I will consider at length the idea that the identity of a causal property is determined by the dispositions it confers on particulars having it. This is one more precise way of articulating the view promoted by semirealism, that realists should understand the causal properties described in our best theories simply in terms of dispositions for relations. This more precise view concerning the nature of causal properties raises several important metaphysical questions, but for the time being let us merely take note of the epistemic fact that regardless of the specific details of the natures of properties, it is only by having some knowledge of the relations in which they stand that one can design and perform experiments to detect or manipulate them.

Thus, for example, it is only by exploiting some knowledge of the relations of the property of negative charge that one is able to design and construct instruments to detect instances of it. It is only by exploiting some knowledge of the relations of properties of cells and their contents that one is able to manipulate enzymes so as to bring about desired rates of biochemical processes. When realists claim to know that a property is instantiated, they do so because the sciences have provided sufficient knowledge (in their estimation) of the relations in which that property stands, and thus the relations of some particulars having it, to render such claims testable and ultimately compelling. This being the case, it is immediately clear that a knowledge of causal properties entails a knowledge of concrete structures. Recall that a concrete structure is a relation between causal properties – a particular kind of relation between particular kinds of properties, families of which are described, in summary form, in the equations of theories. The structures endorsed by sophisticated realists concern causal relations that permit the identification of properties. (And as suggested earlier, this may lead to a knowledge of other things, such as "laws" and principles that summarize features of causal relations.) The identification of these properties requires some knowledge of their causal relations. Thus, a knowledge of properties entails a knowledge of concrete structures.

This is just the beginning, though, for it turns out that this entailment of knowledge goes in both directions! Not only is it the case that a knowledge of properties entails a knowledge of structures, but the reverse is also true. Since, according to semirealism, causal properties are understood simply in terms of dispositions for relations, knowledge of a given concrete structure entails at least some knowledge of the properties that stand in that relation. This is because the very natures of these properties are described, at least in part, in terms of dispositions for that very relation. For example, it is part of the nature of the determinate property, mass 350g, which is (much as I wish it were bigger) instantiated by the slice of pumpkin pie on my desk, that it disposes the slice to accelerate towards the ground were I to be so careless (in an absent-minded contemplation of the properties of tasty pies) as to drop it. This of course does not exhaust the nature of the property mass 350g, which disposes things having it to behave in all sorts of ways, as summarized by various theoretical equations describing the relations of masses to other properties. But any knowledge of concrete structures entails some knowledge, however partial, of the properties whose relations compose those structures.

So, a knowledge of causal properties entails a knowledge of concrete structures, and *vice versa*. This is the bedrock of semirealism, and to the extent that this position represents the evolution of realism today, the bedrock of sophisticated realism *simpliciter*. The prescription this offers for realism, however, requires further qualification, for I have yet to say anything substantial about particulars. With all this talk of properties and relations, one might wonder what has become of the objects, events, and processes one commonly associates with scientific theories. Is there no room for these things in the ontology of the realist? Certainly there is. One reason for not being wholly deflationary about particulars as an ontological category stems from the simple observation that causal properties are not merely distributed in a free-floating or random sort of way across space-time. With high degrees of regularity, they cohere to form interesting units, and the facts of this coherence are what one labels with the term 'particular'. In Chapter 6, I will consider the nature of such groupings of properties in a detailed discussion of the concept of natural kinds. For the moment, though, let me take some care to spell out how a knowledge of properties and structures facilitates a knowledge of particulars. More specifically, for the sake of illustration, let us focus on a specific sub-category of particulars: objects.

Over time, as theories in a given domain are modified and improved, their characterizations of objects may change radically. Even in cases where successive theories ostensibly refer to the same kind of object (the electron, for example), changes in how it is described are sometimes so great that it may not seem entirely credible to maintain that each of these theories takes the same object as its subject matter. How can a realist speak of knowing an object, o, if it turns out that the sciences were and may well now be mistaken in their descriptions of o? Is it reasonable to assert that o exists, and yet be open to the possibility that one's conception of o, in terms of the set of properties a theory associates with it, may change, perhaps greatly? If the meaning of 'o' changes, to what extent is it appropriate to say that one is discussing the same entity? Worries about these sorts of questions fuel concerns regarding the pessimistic induction (PI). If realists are *not* entitled to speak as though one and the same object is the subject of successive theories, given discontinuities in theorizing over time, their commitment to a knowledge of objects seems unwarranted. Even if, by invoking some version of the causal theory of reference, one is able to make a case for the preservation of reference over time, realism courts scepticism if descriptions of objects change so radically that one may, by invoking PI, argue that one is never in a position to know anything substantive about the objects to which one refers, however successfully.

But as we have seen, semirealism offers a response to PI, and this response contains the elements of a realist conception of objects. Imagine that I, convinced the Earth is roughly spherical, agree to a debate with a member of the local Flat Earth Society. It would be taken as given that we both have the same object in mind when speaking of the Earth, despite differences in the properties we attribute to it. The same convention should apply, I suggest, *mutatis mutandis*, to twentieth-century theories of the electron. It is no surprise to the sophisticated realist that a knowledge of certain objects persists across theory change, because semirealism is a form of selective scepticism – it embraces only detection properties, not the auxiliary properties of objects. Detection properties, recall, are the causal properties one knows on the basis of detections. Auxiliary properties, on the other hand, are those described by theories but whose ontological status, whether as causal properties or fictions, cannot be determined on the basis of our causal contact with the world. Once upon a time, realism did not make this discrimination. Thus Michael Devitt (1991, p. 46) describes 'theory realism' as the view that 'science is mostly right, not only about which unobservables exist, *but also about their properties*'. But this is painting with too broad a stroke. A more refined consideration of properties is required in order to alleviate the seeming tension between knowing an object and differing about its properties.

In the last chapter I argued that descriptions of detection properties and concrete structures involving them are likely to be retained as theories develop and change over time. According to semirealism, knowledge of unobservable objects is thus articulated as a knowledge of objects *qua* sets of cohering detection properties, and their relations. Understood this way, objects, like detection properties and concrete structures, are immune to the sorts of worries that give rise to PI (but to a lesser extent, as we shall see). It is reasonable for the realist to speak of an object, *o*, even if it turns out that the sciences were and are now mistaken in their descriptions of it, for the realist believes that some if not all of the auxiliary properties attributed to objects by theories may be attributed mistakenly. It is likewise reasonable to assert that *o* exists while being open to the possibility that theoretical conceptions of it may change, for the realist expects changes in the auxiliary properties associated with *o*, and if all goes well, the attribution of further detection properties. So long as concrete structures are generally preserved, as I have argued realists have good reason to expect, they are generally entitled to claim a knowledge of certain objects, since the relations of their detection properties are in evidence wherever one recognizes their presence, and descriptions of these relations are likely to be retained in successive theories.

The case for admitting objects into the realist inventory, however, is inevitably weaker than the case for admitting detection properties and concrete structures. This is why semirealism is a realism about properties and their relations in the first instance. Particulars such as objects are identified with *cohering sets* of detection properties (whether as mere "bundles" or as "inhering" in substrata, as discussed later in this chapter). Determining whether any given set of properties (such as the mass, charge, and spin of an electron) coheres in the way one expects of a particular is a greater task than merely identifying any one of these properties and some of their characteristic relations. Nevertheless, this greater task is a common part of scientific investigation. It is precisely because the set of properties and concrete structures associated with oxygen is not co-extensive with that previously associated with dephlogisticated air that I earlier suggested that the terms 'oxygen' and 'dephlogisticated air' should not be under-stood as co-referential. There is, I suspect, no clear answer to the question of how many cohering detection properties one must know before it is reasonable to speak of knowing an object. But whatever one says here, it will remain the case that it is easier to be mistaken about objecthood than it is to be mistaken about detection properties and their relations. This is merely a cautionary word to the wise. In circumstances where one *does* have significant evidence for the occurrence of cohering packages of detection properties, the realist has grounds for a knowledge of objects.

Much more could be said about this, for there are various tests of evidential significance that seem appropriate to assessing knowledge claims when it comes to unobservable objects. I will limit myself to one example here. Realists often claim inductive support for their knowledge of objects from demonstrations of corroboration by alternative forms of detection. The greater the extent to which it is possible to detect the coherence of a particular set of causal properties using evidence from different sources, it is argued, the greater the extent to which knowledge claims regarding an unobservable object – as opposed to merely unobservable properties and relations – are warranted. Hacking (1983, p. 201), for example, argues that in microscopy, unobservable objects (in his illustration, dense bodies in red blood platelets) can be detected by different means. The construction and operation of different kinds of microscopes, here, such as light microscopes and transmission electron microscopes, is in each case informed by very different aspects of physical theory. As a consequence, knowledge claims regarding these bodies are all the more reasonable.[1] Conversely, our inability to forge links to the putative causal properties of things like the luminiferous ether, on the basis of which one could detect it and corroborate it, is not unrelated to our conviction that there is, in fact, no such thing.

Let us take stock of the preceding discussion. I have explained how a knowledge of causal properties and a knowledge of concrete structures entail each other, and how this knowledge can be put to work in support of a realist commitment to certain particulars described by scientific theories. The latter point is suggestive in a way that has important repercussions for the evolution of realism. Given conditions under which knowledge claims regarding particulars are epistemically secure, relatively speaking, a knowledge of properties and structures on the one hand, and a knowledge of particulars on the other, are *also* mutually entailed. As I have suggested, in cases where one has grounds for believing that certain detection properties cohere, a knowledge of these properties and the concrete structures associated with them implies a knowledge of particulars. But the entailment also goes in the other direction, for just as one cannot identify a causal property without having at least some knowledge of its characteristic relations, one cannot identify an object without having at least some knowledge of its detection properties and their relations. Thus, where there is evidence of coherence, a knowledge of structures implies a knowledge of particulars.

This is a surprising outcome, and tells us something extremely important about epistemic relations concerning various items that together make up the ontology of the world according to scientific realism. The primary source of inspiration for semirealism comes from a consideration of the selective scepticism of ER and SR. It is thus remarkable that the result should be a position according to which a belief in the existence of certain

---

[1] Salmon 1984, pp. 217–19, and Franklin 1986, pp. 166–8, and 1990, pp. 103–15, give similar examples. Van Fraassen 1985, pp. 297–8, not surprisingly, disputes the significance of corroboration. The fact that different, "theoretically independent" instruments and techniques corroborate one another, he says, offers no support to the realist, because different instruments are intentionally constructed in such a way as to generate similar outputs. The idea of instrumental "fixing" is a familiar topic within the sociology of scientific knowledge.

particulars and a belief in the existence of certain structures entail each other. After all, these earlier approaches to selective scepticism are premised explicitly on the denial of any such implications! ER, recall, advises belief in the existence of certain objects, and scepticism regarding theories more generally, which contain descriptions of their properties and relations. Epistemic SR, conversely, advises belief in structures, and scepticism regarding knowledge of the intrinsic properties and relations of objects which define these structures in the first place. It is thus a striking result that, construed in terms of semirealism, sophisticated forms of realism about particulars and structures entail one another.

Die-hard supporters of ER and SR will not be happy with this. Consider SR, for example. Neither epistemic nor ontic SR is compatible with the conclusion that a knowledge of structures, under certain circumstances, entails a knowledge of objects. According to these versions of realism, theories are correctly interpreted as telling us nothing substantive about properties and particulars that stand in relations. According to epistemic SR, for instance, a knowledge of structures radically underdetermines the natures of entities satisfying them. To the advocates of both epistemic and ontic SR, unobservable entities seem fickle companions next to structures. Entities come and go, but structures are retained across theory change. How could a knowledge of structures that persists throughout successive theories entail a knowledge of entities that are routinely described in vastly different ways over time?

This is a natural question for anyone having the unrefined habit of thinking about particulars in terms of the full complement of detection and auxiliary properties with which they are associated in theories describing them. Having this habit, the naïve realist may view particulars to which theories attribute identical detection properties but different auxiliary properties as *different* particulars. But by now it should be clear that this way of thinking will not do. Continuity across theory change emerges when realists appeal to unchanging attributions of detection properties, whose relations one exploits by means of causal interaction in detections and manipulations. It is these detection properties that particulars must possess, for it is in virtue of these properties that particulars are related to our instruments of detection. Speculation concerning auxiliary properties, on the other hand, is apt to change when theories do, and the sophisticated realist puts no store in them. Thus, concrete structures *do not* underdetermine particulars, but merely their auxiliary properties. And thus, strictly speaking, different ontologies are *not* consistent with the same systems of concrete structures. Structural knowledge implies knowledge of both detection properties and ultimately, by extension, particulars having these properties.

This is not to say, however, that the vast amounts of cognitive energy lavished by scientists on developing the auxiliary contents of theories is wasted. The realist must take care to understand the proper epistemic function of the auxiliary. Though semirealism withholds belief from auxiliary properties, it nonetheless embraces theoretical descriptions that contain them. For these descriptions give rise to further investigations whose aim is to determine whether auxiliary properties are actual causal properties, by detecting them and thereby discovering previously unknown relations. In the process, auxiliary properties are often converted into detection properties or ruled out altogether. The potential heuristic value of auxiliary speculation to the discovery of more accurately described structures should not be underestimated. Auxiliary properties are not themselves objects of knowledge, but they are often methodological catalysts.

Perhaps an example will help to clarify the role of auxiliary content. Worrall (1989) argues that from the perspective of epistemic SR, several of the conceptual problems that confound attempts to understand quantum states (states of entities as described by quantum mechanics) in classical terms cease to be problems at all, because they concern aspects of theories that exceed descriptions of structures. He draws on an analogy to Newton's theory of universal gravitation. Speaking anachronistically, an argument was made at the time of Newton that accepting his theory of gravitation is tantamount to giving up on realism. By accepting the theory, it was claimed, one thereby accepts the unintelligible idea of action at a distance, as opposed to the innocuous idea of contact action, an explanatory principle championed by those inclined towards realism. (Newton's theory offers no mechanism to explain how gravitational forces act over distances between massive bodies.) But this sort of worry, says Worrall, does not trouble the structural realist, who is free of unhealthy desires for a knowledge of underlying metaphysical mechanisms. The epistemic structuralist is content merely to say that Newton and the pioneers of quantum mechanics (QM) discovered certain relations in the world, and described them in terms of mathematical equations. To the extent that they were correct, the structures their theories describe are preserved in later theory. No further explanation is required, or wanted.

This cannot be the whole story, however, for surely part of what drives the sciences forward is attempts to discover and describe new, previously unknown relations and relata, and here the investigation of deeper explanations, in terms of unobservable processes, is often crucial. There is no doubt that many of the auxiliary properties described by present-day theories are fictions. One of the goals of scientific investigation, however, is to uncover grounds with which to discard them, or transform them into detection properties in virtue of which previously unknown relations are brought to light. In the process, the study of auxiliary properties often leads to the detection and corroboration of new particulars and concrete structures involving them. For example, there are quantum mechanical scenarios in which the measurement of a property of one particle can be correlated with the outcomes of measurements on another, even if the measurement events are too distant from one another to be connected causally; such correlations are described as 'non-local'. In QM, studies of so-called hidden variables programs which aim to dissolve the apparently non-local nature of these correlations have demonstrated that certain classes of such programs are untenable. If, in future, empirical tests were designed for the presence of hidden variables, though today this seems the stuff of science fiction, they would be shaped by these early findings. Semirealism neither is too impatient to seek the truth in Newton's theory of gravity or QM, simply because they contain elements one cannot explain in realist terms, nor eschews an interest in deeper questions. Realists should believe in the things for which they have warrant, and let the sciences get on with investigating the rest.

If the arguments of this section are sound, it would appear that ER and SR seriously misrepresent the realist inventory. Their proponents maintain that believing in entities and believing in structures are opposed positions, committed to an exclusive knowledge of different aspects of the world. This opposition is plausible *prima facie*, since entities and structures having to do with the relations of entities are different things. But on more careful reflection the realist finds that entities, such as properties and the particulars that have them, and structures, understood concretely in terms of relations, are epistemically interwoven. They may be ontologically distinct, but so far as one's knowledge of them is concerned, they come as a package. ER and SR are thus incomplete as they stand. The commitments of ER cannot be held in isolation, because they are established on the basis of a knowledge of certain relations. Likewise, the commitments of SR cannot be confined to structures, because on the most promising under-standing of structure, as concrete, structural knowledge yields substantive information about detection properties, and thus about the particulars that have them. Properly construed, a knowledge of particulars and a knowledge of structures fold one into the other. A commitment to one is thus a commitment to both, and for this reason I say they entail each other. They amount to one and the same thing: semirealism.

My listing of the items making up the realist's inventory is now complete. Having made a case for the place of particulars, however, it is time now to consider one last proposal for a realist approach to selective scepticism. We have been through ER and epistemic SR in some detail, but I have left ontic SR until now for a reason. While epistemic SR suggests that realists cannot have any substantive knowledge of the nature of unobservable particulars, such as the objects admired by advocates of ER, ontic SR takes a more radical stand. The central claim of this latter position concerns the issue of whether realists are entitled to believe that objects exist at all! Having clarified what it means, according to semirealism, to have knowledge of particulars such as objects, let us turn our attention to the challenge of ontic SR.

3.3 ONTIC STRUCTURALISM: FAREWELL TO OBJECTS?

SR is the view that to the extent that theories offer true descriptions of the world, they do not tell us about the nature of the unobservable realm. Rather, at best, they tell us about its structure. Until now my discussion of this view has focused on one of its two subspecies, epistemic SR. The epistemic version attempts merely to restrict what can be known about unobservable objects to higher-order, formal properties of their first-order relations, and so nothing regarding the ontology of objects follows from it. As the label suggests, epistemic SR is a purely epistemic thesis. It has nothing to say about the ontological status of unobservable objects aside from the fact that they exist and do, presumably, stand in various relations. Outside of debates concerning scientific realism, there is nothing to prevent a proponent of epistemic SR from investigating deeper questions regarding the basic metaphysical natures of these particulars. In a purely metaphysical context, further removed from the sciences, these people might consider arguments, for example, about whether objects consist in some sort of substrata in which properties inhere, or whether they are rather just bundles of properties, or brute particulars, about which nothing more metaphysically illuminating can be said. But these debates are *a priori* in character, premised on the existence of unobservable objects. Epistemic SR does not foreclose metaphysical possibilities regarding the most fundamental natures of such particulars.

Advocates of ontic SR, on the other hand, hold that realists should believe only in structures described by our best theories, because structure is all there is to reality. This is no mere epistemic claim but an ontological one, with epistemic consequences. The usual talk of objects, they say, is misguided, and engenders fatal metaphysical difficulties. Ontic structuralists are happy to speak of objects, but only as a *façon de parler*, not to be taken literally. This sort of talk must be informed, they suggest, by a more enlightened ontological picture than realism generally assumes. At the core of this picture is the idea that objects, conceived of as bearers of properties that stand in relations, are ontologically otiose. If this is the case, semi-realism, and all other forms of realism according to which one claims a knowledge of objects, are seriously mistaken. In the remainder of this chapter, however, I will argue that the worries motivating ontic SR are insufficient to recommend it, and furthermore, these worries are better addressed in other ways. The precise metaphysics of objects, I believe, remains an open question for the realist, and the idea that they comprise a genuine ontological category is hardly moribund. And as we shall see, fending off the challenge of ontic SR is not merely a negative task, but yields something positive also. By considering it one comes to a deeper understanding of how semirealism views particulars, as a notably heterogeneous kind.

Ontic SR is inspired by certain theories in modern physics, and more specifically, the case for it proceeds from the interpretation of QM. Among its many puzzling features, QM appears to underdetermine the nature of quantum particles such as photons and electrons as regards their identity, or individuality. This underdetermination, claim the advocates of ontic SR, is fatal to any position that incorporates a realism about objects.[2] In the case of observable things and more generally, in classical physics, one thinks of objects as having identities that distinguish them from other objects. Even identical twins, for instance, have separate identities, and, luckily for their parents, can be distinguished from each other, if only because they have different spatio-temporal properties! One would never attribute precisely the same location to both, for example, at any given time. This notion of individuality is reflected in the way one counts objects using classical (or as they are called, Maxwell-Boltzmann) statistics, which recognize different permutations of objects as constituting distinct arrangements. That is to say, if one lines up our friends, the identical twins, and then swaps them around, the pre- and post-swapping states

---

[2] For more details of the physics underlying the claim of underdetermination than I will present here, see French and Redhead 1988, French 1989, van Fraassen 1991, and Huggett 1997. My present concern is with the import of this apparent underdetermination, which is discussed by French 1998, Ladyman 1998, and French and Ladyman 2003.

constitute different arrangements of them. So far as everyday life and classical physics are concerned, the two states represent different physical arrangements of the system of twins.

When it comes to the unobservable denizens of the world of QM, however, one counts objects in a different sort of way, and it is unclear whether quantum statistics respect individuality. Consider two particles of the same kind, such as two electrons. These putative objects have in common all of the state-independent properties identified with their kind, which in this case include particular values of mass, charge, and spin. Now imagine these electrons distributed across two energy states. According to the Standard Model in modern physics, all subatomic particles are classified as belonging to one of two categories: bosons (such as photons), and fermions (such as electrons). But amazingly, neither Bose-Einstein statistics, which pertain to the former, nor Fermi-Dirac statistics, which pertain to the latter, count particle permutations as constituting different arrangements. In the scenario in which each of our two electrons occupies a different energy state, interchanging the particles has no physical significance according to QM. The pre- and post-swapping states do not constitute a different arrangement of the system of electrons.

This suggests that objects described by QM are not objects in the everyday or classical sense, and generates a dilemma concerning their supposed objecthood. One might hold that quantum particles are peculiar individuals – ones that seem to violate the principle of the identity of indiscernibles (PII, famously associated with Leibniz). According to the strictest version of PII, any "two" objects that have all of the same intrinsic properties are, in fact, one and the same object. If it is impossible to discern them as distinct on the basis of their intrinsic properties, they are not distinct. In this strict form, it seems that PII may be violated by everyday or classical objects, since it seems plausible that two distinct things might have all the same intrinsic properties. Perhaps our friends the identical twins could fit this bill, and in any case, whether or not they occur, one can certainly imagine such possibilities. If one weakens PII to include the sharing of relational properties, however, such as spatio-temporal ones, the principle appears to work well with respect to the identities of macroscopic objects in the actual world. A great deal more can be said about how PII might be interpreted and whether it is, ultimately, a reasonable principle, but I will not digress to consider these issues here. Given PII, the important point for present purposes is that unlike classical objects, quantum objects appear to violate even weak versions of the principle, because QM state functions describing assemblies of particles often attribute to them all of the same intrinsic and relational properties.[3]

Thus, if one insists that quantum particles are individuals, one must nevertheless accept that they are peculiar, indiscernible individuals, and their individuality will have to be understood in terms of primitive, individual essences, as opposed to determinate properties. An imponderable essence of this sort is sometimes referred to as a 'haecceity' or 'primitive this-ness'. On the other hand, one might hold that particles are not individuals, but non-individuals of some sort. This interpretation is favoured by some who view talk of "particles" as simply elliptical for talk of excitations of a quantum field. Field quanta, like quantities of water in a glass, are not individual objects *per se*. Proponents of ontic SR argue that these two different understandings of the nature of objecthood are underdetermined by QM. The theory itself does not tell us which interpretation is correct. Let us grant this, for the sake of argument. What is the significance of this underdetermination, assuming it obtains? Steven French and James Ladyman hold that the mere fact of underdetermination scuppers any form of realism involving objects, for as Ladyman (1998, p. 420) contends, 'it is an *ersatz* form of realism that recommends belief in the existence of entities that have such ambiguous metaphysical status'. Ontic SR is immune to this charge, because it makes no recommendation on behalf of objects. It advocates a conception of reality according to which objects are relinquished in favour of structures taken as 'primitive and ontologically subsistent'.

Now, the construction of an object-free ontology is an intriguing metaphysical program, but the argument from underdetermination here gives it little support. *Mere* underdetermination, in fact, suggests nothing at all. In order to motivate the idea that an ontology lacking objects is required, it would help if one had some reason to think that no account of objects, or at least neither of the two options at issue here in the case of quantum objects, is tenable, but mere underdetermination suggests nothing like this. Indeed, some philosophers of physics offer reasons for preferring one of these options to the other.[4] But even if one were to

---

[3] I assume the orthodox interpretation of QM here, for the sake of argument, but there are other interpretations. According to David Bohm, for example, particles have distinct spatiotemporal trajectories, and may thus be regarded as individuals in a more classical sense. Bohm's theory, however, has interesting and problematic idiosyncrasies of its own. For an outline, see Albert 1992, ch. 7, Bohm and Hiley 1993, and Chakravartty 2001.

[4] Redhead and Teller 1992, and Teller 1995 and 2001a, opt for the non-individuals view. Van Fraassen 1991 appears to dispute it. For work on these and related issues, see Huggett 1997 and Castellani 1998 (especially the contributions by Castellani, French, and Teller).

accept this underdetermination as irresolvable, the cure offered by ontic SR – doing without objects – is, I suggest, significantly worse than the disease. Consider the analogy of everyday or classical objects. Here too there is a form of underdetermination. Physics underdetermines the choice, for example, between thinking of these objects as bare substrata instantiating properties, and thinking of them as just bundles of properties. Both views are associated with puzzles: the mysterious nature of the substratum as, in Locke's words, a 'something-I-know-not-what'; the primitive status of the unifying relation, sometimes called 'compresence', that holds bundles of properties together; etc. This has not led us to renounce macroscopic objects, however.

In response, French and Ladyman (2003, pp. 50–1) argue that there is an important disanalogy between the everyday and quantum cases, for observable objects can be experienced 'directly' and identified ostensively (that is, simply by pointing to them). Access to unobservable objects, on the other hand, is unavoidably theoretical, since the construction and operation of the instruments one requires to detect them are informed by theory. And so, if our best theories underdetermine the precise metaphysical natures of these unobservables, this is reason enough to reject them. But this response is no help to ontic SR. Rather, it serves to illuminate an ambiguity in the position, and however this ambiguity is resolved, the disanalogy here between everyday and quantum objects has no bearing on the question of whether metaphysical underdetermination furnishes reasonable grounds for disposing of objects as an ontological category. Let us see why this is so.

The difference suggested here between everyday and quantum objects concerns the fact that the former are observable, and the latter can be detected only with the aid of instruments whose functions are described, in part, by theories concerning the unobservable. Only observable things can be presented ostensively. This, however, raises a question of interpretation for ontic SR: does the position assert that there are *no* objects, or does it rather assert that there are in fact observable objects, but no unobservable ones? Consider the first possibility. If advocates of ontic SR hold that there are no objects *simpliciter*, then it is a strange thing for them to claim that since everyday objects can be ostensively presented and quantum "objects" cannot, one need only dispose of the ontological category of objects in the latter case, but not the former. For on this interpretation of ontic SR, it hardly matters that beach balls and parasols are observable and ostensively presentable – they are no more objects than photons or electrons!

Alternatively, ontic SR might be interpreted as accepting the existence of observable objects, but rejecting the idea of unobservable objects. But then the fact that observable things are ostensively presentable and unobservable things are not turns out to be irrelevant. French and Ladyman suggest that metaphysical underdetermination at the quantum level is more severe than the underdetermination I noted a moment ago, regarding observable objects, because only in the case of the "objects" of QM is the very applicability of the concept of individuality underdetermined, and this, they claim, is a more worrying sort of underdetermination. But observation and ostension *also* underdetermine the status of objects with respect to individuality. Whether or not there are reasonable identity criteria for everyday things like people and bicycles, whether at a time or over time, are challenging metaphysical questions, and the theories of modern physics offer no help with them (if only). Granted, the causal chains that connect unobservable entities to one's senses are generally, but not always, longer and more complex than those required to detect observables. But for the realist this is a matter of degree, not kind, and realists hold that such differences in degree are not by themselves epistemically significant. Whether a putative object is observable or unobservable has no bearing on the question of whether it is properly regarded as an individual.

To think that the disanalogy between observables and unobservables regarding ostensive presentation is important to an assessment of whether the concepts of individuality and objecthood are reasonably applied is to bestow an extraordinary privilege on merely putative objects of sensory experience. Of course, one often takes the things one sees to be individual objects prior to any philosophical reflections about their criteria of identity. But in the context of realism one is concerned with questions of justification, not pre-reflective beliefs. One should not believe in the existence or properties of "objects" experienced in hallucinations or optical illusions, for example, even though these pass the test of ostensive presentation. In order to distinguish between veridical and non-veridical sensory experience, one invokes theoretical beliefs, and as a consequence, the force of the suggested disanalogy between one's 'direct' access to the observable and one's more 'indirect' access to the unobservable is lost. The application of the concept of individuality is theoretical in either case. Ontic SR has not demonstrated that the metaphysical underdetermination by physics of the natures of things at quantum levels of description gives us a reason to jettison objects as an ontological category.

3.4 ONTOLOGICAL THEORY CHANGE

Everyday and scientific discourses are infused with object-talk to such an extent that the contention that it is empty cannot help but seem deeply revolutionary. We quantify, generalize, and perform inductions over objects with such abandon that any picture of reality according to which these practices are metaphysically confused should be required to meet high standards of persuasion. It may be useful to consider here, in more general terms, what sorts of methodological principles might reasonably govern theory change in the area of basic ontology. With these general principles in mind, one would then be in a better position to evaluate whether changes to the realist inventory, like the one promoted by ontic SR, are compelling. In metaphysics there are rough but nevertheless useful guidelines that are commonly brought to bear in deliberations of this kind. Revisions to commonly accepted categories of basic ontology should be considered poorly recommended, I suggest, unless the substitution of one ontological framework for another satisfies at least some of them. Let me summarize these principles as follows:

1 need: there is a fatal or otherwise serious problem with the commonly accepted ontological framework
2 explanatory role: the replacement framework is explanatorily stronger; that is, it serves the same and further explanatory functions
3 primitives: the replacement framework is less obscure; that is, it incorporates fewer primitive notions

The first of these principles concerns reasons one might have to be dissatisfied with the status quo, whereas the second and third are comparative. Switching to the ontological framework suggested by ontic SR, I believe, satisfies none of them, and seeing why this is so will lead to further clarification of how semirealism views the nature of particulars.

Consider first the need principle. I have already argued that the mere fact that QM underdetermines the nature of its objects with respect to individuality is insufficient to recommend the rejection of objects as an ontological category. The situation regarding quantum-level objects is analogous to the situation regarding everyday observables, whose ontological natures, including questions of individuality, are underdetermined by physics. Yet this latter underdetermination does not, it seems, compel us to relinquish everyday objects. The need principle is what proponents of ontic SR appeal to most strongly. Even if they were successful in this appeal, however, the satisfaction of the need principle would not be sufficient, by itself, to recommend the change in basic ontological theory they suggest. This is because the need principle concerns only the rejection of one ontological framework, and the rejection of one does not *ipso facto* recommend another. Ontic SR thus suffers not only from arguments to the effect that it has not made a case for a change in our theory of ontological categories, but also from the fact that satisfying the need principle in isolation would not be convincing in any case, and ontic SR appears to offer no further reason for the change it recommends.

Perhaps there *are* further reasons, though, waiting in the wings. Let us move on to the explanatory role principle. One might be forgiven for thinking initially that the ontological framework suggested by ontic SR is incoherent. This, I believe, is a natural first response. Given that structures are defined by relations, and that relations require relata in order to be instantiated – that is, to be part of the world of the concrete, of interest to scientific realists, as opposed to merely abstract, mathematical entities – ontic SR demands a belief in the existence of concrete relations coupled with a belief in the non-existence of the relata on which they depend, and this is contradictory. This form of relation-relata dependence is a *conceptual dependence*. It is part of the very concept of a concrete relation that it relate *something*. According to our concepts of these things, the former cannot exist without the latter, and in this sense objects play an important, constitutive, explanatory role with respect to our notion of structure. Of course, this explanatory role is not one that ontic SR can be expected to duplicate, on pain of contradiction. And given this, one might argue that this particular explanatory function is not a function that one can reasonably expect ontic SR to serve, since its central claim is a denial of the existence of objects. Ontic SR recommends that one *revise* one's concepts in such a way as to view relations as ontologically subsistent, even in the realm of the concrete. To argue against it on the basis of a violation of conceptual dependence is thus, it seems, question-begging.

There is another form of relation-relata dependence, however, that one might consider in connection with the explanatory role principle, and without begging the question against ontic SR. Let us call this a *causal dependence*. One of the most important explanatory roles served by objects is to provide a means of change. Objects have properties, and it is because they do that things happen to them. As noted earlier, the mathematical equations commonly offered as law statements in the sciences can usually be interpreted as describing relations between the properties of particulars. If one increases the pressure, says Boyle's law, the ratio of the temperature to the volume must increase as a result. If one applies a force, says Newton's second law, the body must experience an acceleration that is proportional to that force and inversely proportional to its mass. Indeed, I suggested the realist should understand these properties simply in terms of dispositions to stand in precisely these sorts of relations. Objects with properties are explanatorily central in the commonly accepted ontological framework. How would an object-less ontology account for change?

Unlike the problem of conceptual dependence, causal dependence represents a non-question-begging difficulty for ontic SR. It is reasonable to expect any account of scientific realism to possess the ontological resources with which to explain how one gets from one state of affairs to another. If one accepts the ontological framework of ontic SR, one appears to be left with explanatory gaps – missing links – between subsequent states of affairs. Given a concrete instance of some set of relations, one has no explanation for what constitutes the active principle that, under the right circumstances, transforms this set of relations into another. In the commonly accepted ontological framework, the "nodes" of structures, occupied by the properties of particulars, are no mere phantoms, posited for heuristic reasons but then relinquished by careful metaphysicians. They have ontological clout. The natures of properties give one something on which to hang explanations of change. If one takes the slogan 'relations without relata' seriously, however, the replacement framework may seem to have insufficient resources with which to provide such explanations. Unable to supply the desired missing links, ontic SR would then run foul of the explanatory role principle.

The argument from causal dependence fares better than the argument from conceptual dependence, but it is not conclusive. Not all versions of the commonly accepted ontological framework recognize the need for active principles of the kind just invoked to provide explanatory links between states of affairs. Those with a taste for desert landscapes, to borrow W. V. O. Quine's redolent image, are happy to analyse events in terms of brute successions of states of objects. Hume, who was prowling in the desert long before Quine journeyed through it, is famous for enshrining scepticism about anything that might be described as a connection between the relata of causation, and offered mere, regular succession instead. Many realists are non-Humean in their appreciation of such landscapes – the terrain is often too barren of explanation for their liking – but this is not always the case. In Chapter 4, I will consider the issue of causation and the idea of "necessary connections" in detail. In the meantime, however, if one admits the option of living in the desert, then an appropriately arid response to the problem of causal dependence is available to the proponent of ontic SR. Causation might be analysed in terms of brute successions of structures, as ontic SR conceives them. Conversely, those who desire explanations for change will find the Humean picture unsatisfactory, whether in its traditional guise, or dressed up in the form of ontic SR.

Perhaps the best reason for thinking that objects are ontologically significant is the empirical discovery, considered earlier in connection with the claim that a knowledge of structures entails a knowledge of particulars, that often instances of certain groups of properties cohere. A particular set of properties come together as a package to constitute an electron, for example, whether one construes this particular as a particle-like object or as an excitation-type event in a quantum field. These sets of properties seem to like each other's company; they are always detected together. Coincidence, or object? Again, it is reasonable to expect a tenable version of scientific realism to offer some explanation for empirical discoveries of this kind. On an object-inclusive ontology, the realist has an explanation for why the potentials for certain types of structural relations always come together: the properties that confer dispositions for these relations cohere in the form of an object. On the ontological framework sketched by ontic SR, however, it is unclear what could serve to provide a parallel explanation.

In fact, this form of challenge to satisfy the explanatory role principle applies not merely in connection with cohering collections of properties (that is, particulars, such as objects), but also in connection with the natures of specific causal properties. A causal property is generally capable of standing in not just one but many different kinds of relations. This fact is easily explained within the ontological framework suggested by semi-realism. Properties are typically "many-faceted" in that they confer dispositions for many different relations, to be manifested in different circumstances, and cohering sets of these dispositions are precisely what one associates with a given property. In virtue of having a mass of 350g, for example, the tasty slice of pumpkin pie I ate earlier was disposed (sadly no longer, given my taste for dessert landscapes) to manifest a host of possible relations, and one associates the collection of these dispositions with one and the same property. The set of dispositions associated with a property is part of, and perhaps all of, what makes a causal property the property that it is. (I will discuss the nature of property identity at length in Chapter 5.) Conversely, on the framework of ontic SR, it is unclear why some of what must be regarded as ontologically insignificant "nodes" occurring in different structures should be identified with one another. Semirealism identifies them easily as the same property. Here too, ontic SR fails to satisfy the explanatory role principle.

Before considering the third and final principle for assessing the desirability of a revolution in ontological theory, the primitives principle, let us remind ourselves of why ontic SR takes objects to be problematic. The fact that QM underdetermines the nature of quantum particles as regards their individuality is taken to cast doubt on more traditional forms of realism. A realism that admits the existence of such metaphysically ambiguous entities, it is claimed, is not worth retaining. In the next and final section of this chapter, however, I will argue that QM does indeed allow for an object-accepting realism. Crucial to this understanding will be the semirealist conviction that scientific realism is properly, first and foremost, a realism about properties. As we shall see, how objects are then "constructed" from properties is open to a degree of classificatory convention. And though properties are amenable to collection into different sorts of objects, and more generally into different sorts of particulars, this in no way undermines the status of objects as composing a genuine ontological category.

3.5 RETURN OF THE MOTLEY PARTICULARS

The worry motivating the rejection of objects by ontic SR concerns the notion of individuality. What does one require of the concept of an individual, or in other words, a particular? A particular is a *unity* in space-time; it is something that coheres, and has a location. Some think that particulars should placate other intuitions as well, and these are widely disputed. I will catalogue these intuitions shortly, but for the time being let us work with this minimal characterization of what it is to be a particular. Semirealism holds that the relata of relations defining structures are properties in the first instance. An immediate question then arises as to which particulars *have* these properties. This question, I submit, can be answered in different ways, and without compromising a realist attitude towards particulars. In the case of QM, ontic SR describes one of the underdetermined options available in terms of individuals that violate PII. On this option, some form of haecceity is required to distinguish the particles. Ontic SR describes the second underdetermined option as invoking 'non-individuals'. Here "particles" are understood as excitations of a quantum field. But note: excitations are events, and events are particulars! That is, they are collections of property instances that cohere at specifiable space-time locations. Far from doing away with individuality, the second underdetermined option collects property instances (that one might otherwise collect together as objects) in a different way. It collects them together as particular events.[5]

The moral here is that however realists choose to construct particulars out of instances of properties, they do so on the basis of a belief in the existence of those properties. That is the bedrock of realism. Property instances lend themselves to different forms of packaging, but as a feature of scientific description, this does not compromise realism with respect to the relevant packages. This is not to say, of course, that there are never reasons for adopting or rejecting some packages or preferring some to others. Appropriate reasons in this context may be empirical, as when specific problem solutions favour specific ways of describing the phenomena, or theoretical, whether the relevant concerns arise from physics or metaphysics. But given that ontic SR seeks to problematize the ontological status of *objects*, here, let us continue for the moment to focus our attention on them. Let us consider the matter of individuality on the commonly accepted ontological framework and see whether the replacement framework suggested by ontic SR is any more compelling with respect to the primitives principle.

It is no small question how, on the accepted framework, different property instances cohere so as to constitute an object. What makes a collection a unity, distinct from others? The nuances of different answers to this question have been carefully articulated over the course of 2,000 years, and I will not attempt to reproduce these nuances here. It is no injustice in summary, however, to say that one can divide these traditional positions on the nature of objecthood into two broad camps, each of which has two main branches. The broad division is between realism about universals and nominalism. Views regarding universals further subdivide into theories that, on the one hand, understand different objects as distinct substrata which instantiate properties, and on the other hand, bundle theories. Nominalism subdivides into traditional varieties, which view distinct objects as further unanalysable, and trope theories, which describe objects as distinct bundles of tropes. The point of listing these different possibilities here is not to consider them in detail, but merely to note certain aspects of them which may be regarded as fundamentally mysterious and unsatisfactory, or further unanalysable and acceptable, depending on the

---

[5] Jeremy Butterfield reminded me in a timely fashion of Davidson's treatment of events as particulars, in part inspired by Quine. See Davidson 1980.

position one adopts. My aim is not to champion any one of these views, so let us refer to these aspects neutrally as 'primitives'.

The unity of an object on the substratum view is conferred by the bare substratum, the 'something-I-know-not-what'. Not only is the nature of this bare particular unknowable in principle, but the nature of the instantiation relation – the manner in which properties inhere in the substratum – defies further analysis. Individuality is understood in terms of individual essences or haecceities possessed by substrata, whether 'thick' (something property-like, though not a property) or 'thin' (brute numerical difference). The bundle theory does away with substrata but supplies another primitive in the relation of compresence, or collocation. Distinct relations of compresence furnish the bundle theory's account of individuality. For the nominalist, particularity is itself a primitive notion, as are resemblances between particulars. Trope theory adopts precisely the same attitude towards property instances, which have a brute particularity, and can be more or less (up to a maximum of exactly) similar to one another. Bundles of tropes, like bundles of properties construed as instances of universals, stand in a primitive relation of compresence, and distinct relations of compresence delimit distinct objects.

Does the ontological framework suggested by ontic SR offer a less obscure set of primitives than the traditional views, just reviewed, of the framework it hopes to replace? It is difficult to see how it could. To the charge that an object-bound realism recommends metaphysically ambiguous entities, one might say: *tu quoque.* Structures are defined by relations, and various of the metaphysical ambiguities that pertain to objects apply to relations also. Perhaps relations are universals, or perhaps they are better understood in terms of nominalism, simply as sets of ordered n-tuples (pairs, triples, and so on). It is unclear how the resemblance of one instance of structure to another, whether in the same lab at a different time or in another lab altogether, is to be analysed if at all. *What constitutes the individuality of an instance of structure?* There had better be an answer to this question, for instances of structure, if they are concrete, are no less particular than the objects ontic SR seeks to replace. The answer to this question, however, is underdetermined by physics. Things subject to empirical investigation naturally raise questions of individuality, whether they are objects, events, or – given the ontological framework of ontic SR – instances of structure. Since the natures of structures are underdetermined by physics, they are no less metaphysically ambiguous than objects. As a recommendation for ontological revolution, ontic SR fails to satisfy the primitives principle.

A very specific form of ontological relativism seems an appropriate attitude for the scientific realist here, regarding attempts to come to grips with the natures of our most basic ontological categories. From a realist perspective, one might well regard these systems as different, basic accounting methods for keeping track of the same mind-independent stuff. In this context, then, pragmatic as opposed to purely epistemic criteria will be important to any comparative assessment of rival ontological frameworks. The principles I described earlier, governing ontological theory change, are examples of the sorts of pragmatic criteria that are likely to feature centrally. How one thinks about particulars is an interesting and important question, but there is no *a priori* reason to think that just one account should apply across the board. The objects described by QM may not be the same *types* of objects as observable ones. Some objects are countable (proteins, cells), but other "objects" are merely quantifiable (quantities of plasma, light), and thus qualify more loosely. Some appear to persist in time, but others may exist only in the context of specific events during which their properties are instantiated. It is likely that the question of how one gets from properties to objects is best answered in different ways, depending on the objects in question. Objects in general comprise a heterogeneous kind.

Quine (1976) speculates as to whether a careful consideration of fundamental ontology will lead ultimately to a view according to which objects wither away, and one is left with nothing more than regions of space-time with properties, or even more radically, with nothing more than mathematical entities. If objects are heterogeneous in the ways I have suggested, however, these views of ontology seem overly deflationary. After all, one does discover that various kinds of property instances regularly cohere in interesting ways, and thus it is only natural that one should acknowledge this fact in one's speculations about ontological categories. Within constraints furnished by nature, one often groups properties conventionally, but this does not render these groupings unreal. Biological species, for example, are not unreal despite the fact that they are demarcated conventionally and can be demarcated in different ways in order to serve different theoretical and explanatory tasks. An innocuous anthropocentrism should not be taken to imply a worrying "unreality". In Chapter 6, I will argue that scientific categories of kinds of objects, events, and processes are delimited in ways that reflect the types of problems these classifications are intended to address, and our interests in addressing them. But so long as scientific classification reflects the coherence of certain groups of causal properties, the resulting categories can certainly be thought of as features of a mind-independent world.

There is another sense in which descriptions of the fundamental natures of things may vary. In addition to the heterogeneity of kinds of objects and the conventionality of kind classification, one's ontological characterizations of *one and the same* particular may vary depending on the sorts of questions one is attempting to answer. Recalling the semi-realist account of properties and particulars, I suggest the reason one is able and in some cases may be required to characterize cohering collections of properties in different ways (for example, electrons-as-particles versus electrons-as-excitations), is that these properties are dispositional. One usually describes dispositions in terms of their manifestations, and particular manifestations occur only in particular kinds of circumstances. That is why different ontological categories may be better suited to describing a particular in different contexts, involving different investigations or problem types. Descriptions of particulars are often given in terms of descriptions of dispositions that are relevant to particular kinds of interactions, measurements, and so on. Often, the very same properties can be investigated in different ways, and the results are best described in terms of different ontological categories of particulars.

At some points ontic structuralists speak as though the revolution they propose in ontology is not so radical after all. Consider the following passage from Ladyman (1998, p. 42):

Objects are picked out by individuating invariants with respect to the transformations relevant to the context. Thus, on this view, elementary particles are just sets of quantities that are invariant under the symmetry groups of particle physics.

Taken in isolation, the view expressed here is compatible with the idea that the natures of subatomic particles can be understood in terms of collections of causal properties ('invariant quantities'). And if this were all that ontic SR is properly construed as saying, there might appear to be little difference between the structural conception of "objects" it proposes, and the conception of objects offered by semirealism.

Given the further commitments of ontic SR, however, even the most irenic comparison must admit at least one very significant difference. French (2003, p. 257) holds that entities 'cannot be regarded as prior to or onto-logically separate from the structure that yields them'. But as I maintained previously in connection with Psillos's remarks on structuralism, there is a real distinction between the natures of entities and structural relations. According to semirealism, the causal properties that one associates with the nature of an entity may be present in circumstances in which various structural relations – manifestations these properties confer dispositions *for* – do not obtain. No doubt *some* relation (or relations) involving a concrete entity obtains at any time during which the entity can be said to exist, but since causal properties generally confer dispositions for many different relations, they are generally present quite independently of whether any one specific disposition happens to be manifesting. Quantities of gas, for example, are not always expanding, though their properties dispose them to do so under certain circumstances, and massive forkfuls of pumpkin pie are not always accelerating towards my mouth, much as I might wish this were so.

There is something important to be learned from SR, but it is not the idea that there is only structure. Rather, it is that relations between things, both observable and unobservable, are of paramount importance in connection with a realist understanding of scientific knowledge. It is only by means of these relations that one learns anything at all. Our knowledge is constrained by the relations of which things are capable. As a consequence, scientific knowledge is primarily *about* these relations and, I would add, the dispositions things have to enter into different kinds of relations in different circumstances. We have come a long way to learn these simple tenets underlying scientific realism today. In the face of challenging antirealist arguments, realism has done well to move in this direction. Its development in response to these evolutionary pressures has resulted, I believe, in the highly adapted and adaptable view I call semirealism: a sophisticated realist approach to well-detected properties, particulars, and concrete structures.

PART II

Metaphysical foundations

CHAPTER 4

Causal realism and causal processes

To many of us who love desert landscapes ... a proliferation of
what there is might appear repugnant. Unfortunately, jungles remain
where they are, whether we like them or not.

Zeno Vendler 1967, p. 704

4.1 CAUSAL CONNECTIONS AND DE RE NECESSITY

I began with a rough, first approximation view of scientific realism:
scientific theories correctly describe the nature of a mind-independent
world. But this turned out to be rather too rough and too approximate,
and as a consequence realists have proceeded to characterize their understandings of scientific knowledge in more plausible ways. They add caveats
to the effect that realism should embrace only theories that are non-*ad hoc*
and sufficiently mature, or that occur in sufficiently mature domains of
theorizing which tend to be non-*ad hoc*, and that theories are often only
approximately true but on the right track, and increasingly so over time.
Even these refinements, however, are insufficient in light of the pessimistic
induction (PI), and among the various attempts to cope with this challenge, the most promising forms of realism exemplify the strategy of
selective scepticism – believing in some but not all aspects of theories, in
accordance with some epistemic principle of demarcation. Entity realism
(ER) and both the epistemic and ontic forms of structural realism (SR) are
helpful but problematic stages on the road to what I described as a further
evolution of realism. Any form of selective scepticism is *ipso facto* a form of
selective *optimism*, and in semirealism the realist has reason for both. That
was the story of Part I. Semirealism, I believe, offers an understanding of
what a plausibly defensible scientific realism can be.

Recall that speculative metaphysics, broadly construed, is a project that is
interested in and that actively pursues explanations of aspects of the world in
terms of things inaccessible to the unaided senses. As I have described it,

realism commits to a knowledge of various unobservable properties and relations, which can be described as concrete structures, and on this basis to a knowledge of various particulars, including unobservable objects, events, and processes. In virtue of claiming such knowledge the realist is thus already a metaphysician. Within speculative metaphysics, however, there are degrees, and some would urge the realist to stop here, in the shallow end. If one were to refrain at this stage from saying anything more of a metaphysical nature, perhaps even empiricist critics would find themselves able to tolerate if not excuse realist commitments to unobservables. After all, claims regarding unobservable structures are generated in the course of empirical investigation, and this proximity to experience might mitigate the unreasonableness of these claims, to some extent, in the eyes of empiricist critics. Scientific practice, with some exceptions, is conducted at a significant distance from the armchair, where *a priori* speculation is king. Thus, some realists may be tempted to stop at Part I.

On the other hand, I suspect that many will wish to carry on. Throughout my sketch of the recent history of realism, I have made recourse to things whose natures cannot be illuminated by investigations in a purely empirical context. One's concepts of these things are informed by empirical data, no doubt, and must be consistent with them, but these concepts cannot be clarified by empirical investigation alone, nor indeed much at all in the absence of deeper metaphysical consideration. For example, the epistemic engine that drives realist beliefs regarding certain unobservable properties, structures, and particulars is causation. The realist claims that given circumstances in which one appears to engage in significant causal interactions with the objects of one's beliefs, there are grounds for realism, and in such a way as to account for both optimistic and pessimistic inductions on the history of the sciences. The nature of causation, however, is not something that can be settled by empirical investigation alone.

Here the practice of science and the art of being a scientific realist may well diverge. Many different positions including realism and a variety of antirealisms are consistent with the practice of science, and this includes positions that reject metaphysical speculation as a means to belief (if not scientific theorizing). Being a realist in connection with scientific knowledge, on the other hand, raises questions not only about the unobservables explicitly described by scientific theories, like neutrinos and mitochondria, but also about the unobservables realists believe to be *implicitly* described, and that support knowledge claims regarding the former. These latter features of reality, described explicitly in metaphysics, are the foundational supports of semirealism. Most importantly, they include the nature of causation and the idea of natural kinds. Philosophers who give accounts of realism often say little or nothing about these things but then proceed to rest their realism squarely on them, as assumptions to be accepted *ex cathedra* or as promissory notes. Conversely, in the next few chapters, I will address these topics directly.

The motivation for considering such foundational issues deserves further comment, however, for it is by no means obvious that scientific realists *must* address these issues in order to furnish promising descriptions of realism. So long as there *are* tenable accounts of things like causation and kinds, is it not the case that any will do? Indeed, one reason for the diversity of formulations of realism one finds in the philosophical literature is that different realists often have different conceptions of the metaphysical underpinnings of their position, embracing everything from empiricist views to more elaborate metaphysical theories in these domains.

There are two principal reasons, however, why I think pursuing these questions in connection with realism is an important exercise. Firstly, to recall the terminology of van Fraassen's theory of epistemology from Chapter 1, my aim is to demonstrate that realism is an internally consistent and coherent stance. To say that having tenable accounts of the metaphysical constituents of one's realism would help to make it coherent is not the same thing as actually having such accounts, and it is this latter, more ambitious goal that interests me here. Are the foundations hollow, or is there something to them, supporting the edifice of Part I as the realist assumes? Secondly, I believe the metaphysical proposals I will outline display an appealing unity: the notions of causal processes, *de re* necessity, and scientific classification developed here are integrated with one another, and with the ontological commitments to properties, particulars, and concrete structures canvassed in Part I. I will not argue for the exclusive coherence of the metaphysical account I propose, however; indeed, I suspect that others may be possible. More and less elaborate metaphysical foundations are generally correlated with more and less room for explanation, respectively, and I will attempt to clarify certain trade-offs between ontological commitment and explanation in the pages that follow. Let us turn now to the deeper, metaphysical foundations of scientific realism.

Suspense is hard on the nerves, so let me give some indication of what is to come. I will argue that there are, in fact, accounts of causation and natural kinds that are not only compelling on their own merits, but that also fit naturally with one another and provide an excellent base for the commitments of semirealism. Furthermore, one of the nicest features of these accounts is that a helpful understanding of the concept of a law of nature, also commonly invoked by realists, neatly emerges. Perhaps surprisingly, I will suggest that these views go little or no further in violating the strictures of empiricism than I have gone already. On the proposals to follow, I believe that the metaphysical foundations of semirealism are relatively modest. Indeed, one of my motivations in elaborating them is to shed some light on the fact that many realists appeal to more than they need, resulting in versions of realism that are open to criticisms they would otherwise avoid were it not for their excess, problematic baggage.

With these goals in mind, let us begin by considering the nature of causation. Semirealism extracts from ER the epistemic warrant associated with one's ability to interact with and manipulate unobservable entities. This furnishes a criterion for ordering unobservables along a spectrum of entities described by theories, with properties and particulars about which realists are fairly certain at one end, and those about which they are fairly sceptical at the other. It also allows them to distinguish the concrete structures described by theories to which they should commit, from those that belong to their auxiliary content. How should the realist think about this phenomenon, causation, on which so much depends? Philosophy is home to numerous theories of causation, ranging from metaphysically lightweight or deflationary views, which redescribe causal talk in terms of non-causal categories of things, to metaphysically weighty views, according to which causation is an undeflatable *sui generis* element of reality. Indeed, theorists about causation agree on precious little. In seeking a place to start, then, perhaps the realist should begin with as little as possible. Let us begin with the denial of causation.

In spite of the many different views they hold, philosophers interested in causation are unanimous about one thing. All are amused by the inflammatory introduction to Russell's (1953/1918) early paper 'On the Notion of Cause', in which he claims that the idea of causation, 'like much that passes muster among philosophers, is a relic of a bygone age, surviving, like the monarchy, only because it is erroneously supposed to do no harm'. This shared comic relief, however, masks a common misrepresentation of Russell's position. It is often suggested in discussions of causation that the primary motivation behind Russell's attack is the idea that the very notion of it has no place in the sciences, but this is plainly not what Russell believes. Elsewhere he considers causal phenomena in some detail, and is at pains in later work (for example, 1948) to give an analysis of these phenomena in terms of what he calls 'causal lines' – entities that persist along spatio-temporal trajectories. This is no change of heart. In fact, these different parts of the Russellian corpus are consistent with one another. His target in 'On the Notion of Cause' is not causation *simpliciter*, but rather what he refers to as 'the law of causality', or in other words, the philosopher's notion of causation. It is this putatively philosophical creation for which he thinks the sciences have no use. One aspect of this creation is what I will refer to as 'causal realism'.

Realism about causation requires two things. Firstly, according to the causal realist, causation is objective, meaning that it is something that occurs in a mind-independent, external world, as opposed to something that is merely subjective, a feature of one's thoughts or perceptions alone (that is, merely an idea or a concept). The distinction between objective and subjective causation is thus focused on the issue of mind-independence. Secondly, according to the causal realist, causation involves some sort of necessity with respect to the connection between causes and effects. Russell adverts to this aspect of causal realism when he uses the phrase 'law of causality' to describe the view he opposes. Though some dispute that it should, the idea of a law usually connotes some form of necessity. What manner of necessity is a question answered in different ways by different causal realists, but all sign up to it in some form or other. I will return to this question in section 4.4 and in Chapter 5. In the meantime, let it suffice to say that the label 'necessity' is intended to indicate the view that there is more to causation than mere constant (or probabilistic) conjunctions of events. Subjective accounts of causation hold that if there is such a thing as causal necessity, it is an idea or a concept only. Objective accounts hold that if there is such a thing, it is a feature of the world quite apart from our ideas or concepts. In the jargon, causal realists subscribe to necessity in the world, or *de re*, as opposed to mere necessity *de dicto*, which pertains to what is said or thought.

The two most celebrated rejections of causal realism are customarily attributed to Hume and Kant, respectively. The Humean tradition rejects the idea that instances of what one refers to as causation should be identified with anything more metaphysically noteworthy than simple successions of events, and describes the notion of causal necessity as an idea, a figment of one's psychology, that many are tempted to project, but without warrant, on to nature. Hume famously denied one route to a belief in objective causal necessity – the route of observation – and he was of course correct, for one does not have sensory experiences ('impressions', to use Hume's terminology) of necessary connections. Some dispute this, however, arguing that one does have such impressions in experiences of bodily forces. Evan Fales (1990, pp. 11–25), for example, notes suggestively that when one engages in activities involving pushing or pulling or being pushed or pulled, one experiences sensations of force or power, and Hume (1975/1777, p. 67) makes similar observations regarding impressions of bodily force or exertion ('nisus'). But instances of many causal phenomena offer no such impressions, and more importantly, impressions of force are irrelevant here, simply because they are not tantamount to impressions of necessary connections. Forces and causal necessity are different things, and only the former are amenable to sensation. Causal necessity is unobservable, and unlike many scientific unobservables, it is not even a possible object of detection.[1]

Scientific realists need not be causal realists. They may, for example, adopt a Humean approach and thereby remain aloof from further discussion of the nature of causation. Many scientific realists are also causal realists, however, and the reason for this can be traced to the inspiration they take from what I earlier described as the metaphysical stance. As a matter of epistemic principle, scientific realists are interested in and pursue explanations for many observable phenomena in terms of the unobservable things that underlie them. This epistemic commitment yields a knowledge of detectables such as neutrinos and mitochondria, but is also taken by many to promote investigations into certain undetectables, and causal necessity is a case in point. The idea here is not that scientific realists are obligated to be causal realists, but rather that, given their epistemological predilections, it should come as no surprise that many are. Here as in other cases we will encounter later, the motivation for exceeding a Humean metaphysic in favour of something more elaborate is tied to the goal of explaining the phenomena more fully.

Thus, the mere existence of regular patterns of events including the regular behaviours of particulars may be sufficient for purposes of prediction, but, given the central role of causation as the epistemic engine of semirealism, it is not surprising that many realists are apt to wonder why such patterns exist at all. The notion of objective causal necessity, if tenable, serves an extremely important explanatory function for those apt to wonder: it allows the realist to distinguish between the causal regularities

---

[1] For a description of causal realism in similar terms, see Costa 1989, pp. 172–4. Here I present the standard view of Hume, but some argue for the surprising thesis that Hume was a causal realist; see Craig 1987 and 2000 and Strawson 1989. Despite his discussion of bodily forces, Fales 1990, p. 317, n. 26, agrees that one does not perceive causal necessity, because one never has impressions of the totality of causally sufficient conditions for effects. This is a *non sequitur*, though, for even if one had such impressions, they would not amount to impressions of necessary connections.

on which the detections and manipulations of semirealism depend, and merely accidental series of happenings. Philosophers differ on the question of whether such explanations are required or desirable. Those who adopt the empirical stance at this juncture think not, and others disagree. Momentarily, however, I will be concerned with a related but different question. Is the notion of causal realism a live option for those who would subscribe to it? I will contend that despite arguments to the contrary, there is indeed a tenable view of objective causal necessity.

Turning now from Hume to Kant, an appeal to the epistemological sensibilities of scientific realists more generally may help to explain why many semirealists are unlikely to adopt a Kantian scepticism regarding causal realism. On Kant's view causation is again understood in part psychologically, as part of a basic conceptual apparatus (the 'categories of the understanding') that humans possess *a priori*, required for the very possibility of experience. It is simply a bad question on this account to ask whether such concepts apply to the noumena, the mind-independent 'things-in-themselves', for these things are beyond the remit of what one makes intelligible by means of the categories. For Kant, the concept of causality is required for the judgment that an event has taken place. It is by applying the 'rule of cause and effect' that one experiences causal phenomena. Causal necessity is thus a purely representational aspect of things, not a feature of a mind-independent world. For this and other, quite general reasons, this picture is alien to the realist perspective. For example, the mysterious dependence of the phenomena, which can be known, on the noumena, which are unknowable, may seem problematic to the scientific realist. Since causation is understood in terms of an application of the categories, the dependence is not causal, and as a result the noumena are robbed of the ontological and explanatory roles they might otherwise fill. Lacking a role, things-in-themselves are prime targets for scepticism. From a realist perspective, with no grip on the noumena, Kant's transcendental idealism runs the risk of collapsing into idealism *simpliciter*.

It should be clear, I hope, that the preceding remarks are not at all intended as refutations of Hume or Kant, but rather as indicating possible reasons for unease concerning their approaches. The Humean and Kantian rejections of causal realism have been widely discussed, and it is not my intention to add to those discussions here. Nevertheless, for the reasons I have suggested, many scientific realists are motivated to resist these accounts and to investigate the possibility of causal realism instead. Leaving the objections of Humean empiricism and transcendental idealism aside, however, a third class of objections remains. Unlike the views just mentioned, emanating from Hume and Kant, these latter arguments cannot be set aside on the basis of a conflict with the epistemological commitments of various scientific realists. This third category takes issue with causal realism, not by adopting what can be regarded as incompatible positions *ex ante*, but rather by attacking causal realists on their own terms, and thus must be confronted head on. Starting from the assertion of causal realism, these arguments go on to suggest that it is incoherent *by its own lights*. Perhaps having an incoherent idea or even an incoherent category of the understanding might not be intolerable. After all, as epistemic agents, human beings are less than perfect. But if causal realism requires that one attribute an incoherent phenomenon to a mind-independent world, clearly causal realists are in trouble. This line of argument was pressed by the sceptics of ancient Greece, and this very same worry is what Russell exploits in his attack on the philosopher's 'law of causality'. It is this worry that I will attempt to spell out, consider, and dissolve in the rest of this chapter.

4.2 IS CAUSAL REALISM INCOHERENT?

Arguments for the incoherence of causal realism have a venerable history in western philosophy, going back at least as far as Sextus Empiricus (Mates 1996, pp. 175–7) almost 2,000 years ago. Russell's (1953/1918) formulation is something of a twentieth-century *locus classicus*, and his own later account of causation directly inspires several more recent and influential views of causal processes, which I will consider in section 4.5. His critical arguments in 'On the Notion of Cause' have two main targets: the idea of causal necessity (pp. 174, 177, 183), and the Humean definition of causation in terms of constant conjunctions of events (pp. 174–5). The principal arguments for incoherence arise in the latter context but quickly become relevant to the former (pp. 177, 183), which is my main concern here. There are several related lines of argument to consider in this connection, and in order to keep things straight it may help to know where we are going. I will begin by outlining generalized versions of the arguments for incoherence. Next, I will argue that the traditional account of causal realism, according to which causation is a relation between events, fails to offer a compelling response. Finally, I will propose a different account of causation, a "process" theory, and suggest that not only is it immune to worries about incoherence, but that it facilitates various answers to the question of how realists might best understand *de re* necessity in the context of causation, including the one I will go on to defend in Chapter 5.

First, then, let us consider the arguments for the incoherence of causal realism. Imagine a scenario in which one event, *A*, putatively causes another event, *B*. There are three potential problems here, and I will refer to them as follows:

1 the contiguity objection
2 the regress objection
3 the demand for a causal mechanism

In each of these objections, as we shall see, it is the core assumption of causal realism – that of *de re* necessity – that is crucial in eliciting what appear to be problematic consequences.

The contiguity objection

In order that *A* cause *B*, *A* and *B* must be contiguous in time. (One might also consider a variation of this argument, beginning with the assumption that causally related events must be spatially contiguous as well. I will leave aside issues concerning the spatial demarcation of events and focus on the temporal case here.) Of course, one often refers to events as causally related even though they are not temporally contiguous. Knocking over the first domino in a sequence is a cause of the falling over of the last one, for example, despite the fact that the events of the first and last fall are not contiguous in time. But in these cases one also says that there is a causal chain of events connecting the ones originally cited as causally related. That is, one appeals to intermediate causes and effects, and holds that each link in the chain is temporally contiguous with the next. So for *A* to bring about *B* causally, not mediated by other events but directly, *A* and *B* must be contiguous in time. But *A* and *B* cannot be contiguous, because time is dense. In other words, between any two instants, say that at which *A* terminates and that at which *B* begins, there are always further instants. Therefore, it is impossible for successive events to be temporally contiguous. Thus, *A* cannot cause *B*.

The contiguity objection is a *reductio ad absurdum* argument. It begins by assuming a scenario in which *A* causes *B*, and ends with the conclusion that *A* cannot cause *B*. The force of this *reductio* rests on two explicit premises. The first is that causally related events must be temporally contiguous, and the second is that time is dense. Both of these premises seem well founded and, at the very least, hard to resist. Regarding the latter, to be fair, some have speculated that time may be quantized and thus not dense at the Planck scale, which is described by a system of extraordinarily small units of certain physical properties and quantities (the unit of Planck time, for instance, is equal to 5.3906×10^{-44} seconds). However, the idea that time is continuous, which implies that it is dense, unbroken, and infinitely divisible, is widely assumed for a variety of reasons. For example, topological studies of dimensionality (Hurewicz and Wallman 1969/1948, chs. 1 and 2) suggest that non-continuous, discrete time would have dimension zero, which conflicts with the standard assumption that time is one-dimensional. The assumption of continuous time is also helpful in a number of explanatory contexts. One thinks of properties such as velocities and accelerations, for example, as instantaneously possessed by particulars, and assuming that time is dense allows one to define these properties by taking their limit values at an instant.

Add to these benefits the fact that the idea that time is dense also helps in dealing with otherwise disturbing puzzles. Consider for example Zeno's paradox of the arrow, which he offered as a proof for the impossibility of motion. Imagine an arrow in flight. It occupies a space with length no greater than its own at every temporal instant throughout its journey; an instant is a point in time, and has no duration in which the arrow could move. If time is not continuous, how would one explain the manner in which the arrow gets from the space it occupies at any one instant to the space it occupies at the next, given that there is no time between them? It seems impossible for the arrow to make the jump, so to speak. On the other hand, if time is continuous and thus dense, the very concept of a next instant in time is incoherent, because there are always instants between any two. And so, assuming that time is dense, the paradox of the arrow evaporates. On this assumption there is nothing more to the motion of an arrow than the fact that it occupies a space at every instant contained within the duration of its flight.

The second premise adopted by the contiguity objection, that of contiguity itself, is especially difficult for the causal realist to deny. If events are the relata of causation, it is difficult to see how their connection could be necessary if they are not contiguous. For if a putative cause A is not in fact contiguous with its effect B, it is always possible that something might intervene subsequent to A so as to prevent B. That is why the contiguity objection targets direct as opposed to mediated causation, for it is usually assumed by causal realists that within a causal chain of events, relations of necessity generally obtain only between contiguous events, not between events separated by others in the chain. Here we have a first glimpse of the ambiguity of the term 'necessity'. In different contexts the language of necessity is used to express different kinds of metaphysical facts. In some cases, by describing something as necessary, one means that it is a necessary condition of something else. In other cases one means that it is a sufficient condition, or in other words, that it necessitates something. In the context of the contiguity objection, necessity means sufficiency. It is assumed that *A* is sufficient to bring about *B*, and that is why they must be contiguous. I will return to this point momentarily, but first let me continue the outline of the arguments for the incoherence of causal realism.

The regress objection

Shifting the focus from the connection between the putative cause-and-effect events, *A* and *B*, a second challenge aims at connections between the temporal parts of *A* and *B* themselves. Events comprise changes in an object or objects, and given that this is so, there are, presumably, causal relations between their earlier and later parts. (Events are commonly described in terms of change, but this is not always the case. Some also admit unchanging instantiations of properties as events. I will proceed here with the common characterization of events as changes for the sake of argument, since many if not most of what are generally regarded as causes and effects are indeed changes.) If there is a causal relation between the earlier and later parts of *A* or *B*, however, one has the makings of a regress. For if an earlier part of *A* is the cause of the part remaining, then the earlier part cannot serve as the proximate (direct, unmediated) cause of *B*. Only the later part of *A* can serve this role. Whatever duration of *A* one puts forward as containing the proximate cause of *B*, however, one may then ask about the causal relation between the earlier and later parts of the occurrent change in *this* duration. In the hope of arriving at a proximate event to serve as a cause, one may diminish the originally supposed cause, *A*, without limit. The same sort of difficulty applies to the identification of the proximate effect, *B*, of *A*.

One strategy for responding to these difficulties would be to deny the premise that time is infinitely divisible, but infinite divisibility is entailed by continuous time, and as I have observed, there appear to be good reasons for assuming a continuum. And once again it seems that the causal realist's commitment to *de re* necessity is what fuels the argument for incoherence, for if one does not view the connection between causes and effects as necessary, there is no obvious need for a workable understanding of proximate causation. In order to see why this is so, let us return to the issue of the inherent ambiguity of 'necessity', for there is an important difference in the conceptions of necessity at work in the contiguity and regress objections. As noted a moment ago, the contiguity objection derives its force from the causal realist's assumption of necessity understood in terms of sufficiency. The objection preys on the worry that if A and B are not contiguous, there can be no guarantee that A is causally sufficient, given the possibility of intervening events preventing B. It is thus inappropriate to describe the sequence as necessary, since B might not occur following A.

The regress objection, on the other hand, is fuelled by an assumption of necessity that is best understood not in terms of sufficiency, but rather in terms of necessary conditions. David Lewis (1973, p. 563) has this assumption in mind when he argues that 'causal dependence' – the dependence of B on A in cases where if A did not happen, B would not happen either – is not transitive. That is, to paraphrase Lewis, if A causes B, and B causes C, A is a mediated cause of C, but this does not entail that C is causally dependent on A, since it is not always the case that A is a necessary condition for C. Generally speaking, C might have come about in some way other than via A. It is precisely this sort of consideration that leads to the regress objection, since often, presumably, it is only that part of a cause that is temporally proximate to its effect, as opposed to its earlier parts, that constitutes a necessary condition for the effect. Furthermore, presumably it is often the case that causes furnish necessary conditions for their proximate effects only, not the temporally more distant parts of effects. Lewis's analysis of causation is controversial (see Horwich 1987, pp. 170–2, for example), but nevertheless useful in illustrating the intuition targeted by the regress objection. It is the hope of finding unmediated causes and effects, to which one might think the necessity of causal dependence applies, that leaves the realist susceptible to a regress attempting to dismiss all parts of A and B except those serving as proximate cause and effect.

Here is another illustration of the sense of necessity exploited by the regress objection, this time by way of an analogy. J. L. Mackie (1965) famously analyses causation in terms of what he calls 'INUS' conditions. All of the conditions that are jointly sufficient to bring about an effect B, he says, together comprise a complex condition, which is itself generally unnecessary, given that B might have come about in some other way. But when one cites a cause, one is usually interested in picking out some component part of this complex, viz. a part that is necessary to bring about B, but that is insufficient on its own. In other words, causes are *I*nsufficient but *N*ecessary parts of *S*ufficient but *U*nnecessary complex conditions. Consider Mackie's 'N'. When one identifies a cause, he says, one is picking out some part of the conditions preceding B that is necessary. My interest here is not in whether Mackie's account is ultimately tenable, but rather to further illustrate the intuition he illuminates so clearly, that causation involves necessary conditions. The regress objection takes advantage of an analogous intuition on the part of causal realists, focusing on the temporal parts of causes and effects. Realists are invited to associate causal necessity with the temporal part of A that is necessary to bring about B. In searching for that part, however, the causal realist is carried along on a hopeless quest for proximate causes and effects.

The demand for a causal mechanism

The contiguity and regress objections both exploit the idea that time is continuous, thus entailing that it is dense and infinitely divisible. A third and final objection, however, represents a more general concern, and does not depend on any assumptions regarding the nature of time. The concern is general, but as a specific means of introduction let us consider a scenario to which the regress objection does not apply. Imagine a case in which somehow, *per impossibile*, one is able to diminish the event that is to serve as the proximate cause, A, to such an extent that it no longer contains any change between its earlier and later parts. More generally, let us consider not a change at all, but rather some static, unchanging state of affairs preceding B. One might then ask, if A is no longer any sort of change, how does it give rise to an effect? How is it that something static should suddenly bring about a change, when it itself has no element of change within it? How can something that is not a change *bring about* anything? Questions like these are symptomatic of a more general concern about what is happening, precisely, when one thing is thought to give rise to another causally. On further reflection it seems clear that this worry is not limited to cases in which effects are imagined to follow from static states of affairs, but applies to any causal succession. The worrier is seeking some sort of mechanism for the connection between causes and effects, on which to hang the idea of causal necessity.

Historically, the challenge to furnish a description of the nature of this connection has failed to elicit any detailed response. Metaphors abound: links; chains; ties; glue; cement; bringing things about; and perhaps most highly scorned of all, the "powers" of ancient metaphysics which "give rise" to subsequent phenomena. These metaphors, it is maintained, give no useful purchase on the idea of a causal connection. If there is something to the nature of objective causation over and above mere regularity, the realist has yet to furnish anything like a helpful, qualitative description of it. Empiricist critiques of causal realism often contend that if a realist account of causation can offer no description of the nature of a *sui generis* mechanism in which causal necessity inheres, then causal realism is empty. As Simon Blackburn (1993, p. 103) suggests as an exegesis of Hume, 'nothing will do just as well as something about which nothing can be said'. One has no decent conception of what a causal connection *is*, so if this is what the idea of objective causal necessity amounts to, ultimately, one gains nothing worth having by postulating it.

The contiguity objection, the regress objection, and the demand for a causal mechanism offer challenges to the coherence of causal realism on its own terms. They are not objections based solely on Humean scepticism regarding things not amenable to sensation, nor Kantian scepticism based on the status of causation as determined by the categories of the understanding. Rather, they assume the standard realist picture of causation as a mind-independent relation between events, imbued with some form of necessity, and offer *reductio ad absurdum* arguments in the case of the contiguity and regress objections, and an unanswered challenge in the case of the demand for a causal mechanism. On the account of causation these objections assume, their challenges are not easily met. Soon, however, I will suggest an account of causation on which the objections of contiguity and regress do not arise. This view also, though only partially, addresses the demand for a causal mechanism. I will argue that a complete response is beyond the reach of any account of causation, in principle, and that this offers no impediment to causal realism in particular.

The account I will give, however, is not the one causal realists traditionally adopt. In the face of arguments for the incoherence of causal realism, the traditional response of realists has been to stand their ground. But I believe the ground of this response is unstable, and offers no place from which to resist the charge of incoherence. Though the expressions one most commonly uses to describe causal phenomena make reference to relations between events or states of affairs, I believe it is a mistake to think that these descriptions transparently reflect the metaphysical details of causation. In making this mistake, causal realists have left aside other details that would otherwise provide a response to the charge of incoherence. As we shall see, however, causal realists can do better.

4.3 A FIRST ANSWER: RELATIONS BETWEEN EVENTS

If incoherence-type arguments – particularly the objections of contiguity and regress – appear damaging to the prospects of causal realism, it is only,

I think, because everyday descriptions of causal phenomena are ambiguous with respect to the precise details of causation. This ambiguity finds a home in the common practice of identifying events as the principal actors in the analysis of causal relations. Not all realists, however, would agree. The standard response to arguments for incoherence has been to claim that they can, in fact, be answered on an events-based account of causation. In this section I will consider the three responses often suggested by causal realists. I will argue that the first either offers no response at all, or can be reconstructed on a more charitable interpretation along the lines of the second. The second and third responses are suggestive, but on further consideration I will maintain that their promise leads causal realists to a view different from the one to which they traditionally commit.

Consider the contiguity objection. The first strategy for dealing with the charge of incoherence on an events-based account is to contest the idea that there is anything problematic in saying that a putative cause event, A, and its putative effect, B, are contiguous. Beauchamp and Rosenberg (1981), for example, suggest this in a discussion of Russell's formulation of the arguments for incoherence (in defence of their Humean account of causation). *Contra* Russell, they argue, two events can be 'both contiguous and successive if the first begins at instant t_1 and ends at instant t_2, while the second begins at t_2' (p. 196). I suspect that many causal realists take something like this for granted, but on second thoughts it seems clear that this response cannot resist the contiguity objection as it stands, and requires further elaboration. To say merely that t_2 marks a point in time at which A ends and B begins is potentially misleading, because it trades on the ambiguity of beginnings and endings.

This ambiguity emerges when one thinks more carefully about the contiguity objection. The argument here assumes that events are discrete, meaning that they can be defined on closed temporal intervals. Consider a series of instants, ordered in time: t_1, t_2, t_3 ... To say that a putative cause event, A, for example, is defined on the closed interval $[t_1, t_2]$ is to say that A contains (includes) both instants t_1 and t_2, but none before t_1 or after t_2. It is immediately apparent, therefore, why causal realists owe their critics something more precise, for if both A and B contain t_2 the events overlap, in which case they are not strictly successive. If only one of A or B, or neither A nor B, contains t_2, the events cannot be contiguous, given that events are discrete and time is dense, for there will always be instants between them. What, then, is intended by the causal realist when she claims that A ends at t_2 and that this is when B begins? The two possibilities I have outlined here appear to be exhaustive, which might indicate to the uncharitable that this first answer to the contiguity objection is no response at all.

Perhaps this is too quick. Perhaps there is something else the realist might intend here. While the two scenarios I have just described are exhaustive given the assumptions of the contiguity objection, the causal realist might be able to reformulate this first response by rejecting one or more of these assumptions. More specifically, the premise that events are discrete entities invites further scrutiny, and it is precisely this line of inquiry that motivates the second realist response. Imagine a putative cause, A, not on the model of causal relata presupposed by the contiguity objection, but rather on a continuous model. That is, let us define A not as a discrete entity but as a continuous entity. For example, rather than defining A on the closed interval $[t_1, t_2]$, one might instead define it on the half-open interval $[t_1, t_2)$. This would mean that A contains all instants from and including t_1, up to but *not including* t_2. B, the putative effect of A, could then be defined on the interval $[t_2, t_3)$, and so on. In this way, causally related events could be understood as successive *and* contiguous, since there are no instants in time between A and B.[2]

This second causal realist response to the arguments for incoherence is, I think, on the right track. The key to answering incoherence-type objections is to understand causation not as a relation between discrete entities, but rather by in some way appealing to properties of the continuum. The specific attempt to invoke this strategy in terms of the realist's second response, however – reconceiving events in a continuous manner – does not go far enough. There are two difficulties with this proposal that push the causal realist in the direction of a different account of causation. Neither of these difficulties, in my estimation, represents a knock-down argument against the traditional causal realist picture, but they do significantly undermine the idea that the realist should hang her commitment to objective causal necessity on relations between events. Let us consider these points in turn.

The first problem with the causal realist's second response is simply that, while defining events on continuous intervals may counter the contiguity objection, it offers no help with the regress objection. The latter, recall, argues for the impossibility of there being proximate causes and effects, given that any such candidates may be diminished *ad infinitum*. Some, like

---

[2] This sort of approach is taken, for example, by Mellor 1995, pp. 54–6, 219–20. Strictly speaking, Mellor holds that 'facta', not events, are the relata of causation, but his general approach is the same as that described here.

D. H. Mellor (1995; cf. Fales 1990, p. 131), may be willing to accept this consequence. Perhaps causation, like time, is dense, and there is no such thing as proximate causation. If causal realists go down this route, however, they may have to give up causal necessity in the form of necessary conditions as part of their general account of causal realism, for as discussed earlier in connection with Lewis, one may doubt whether this form of necessity applies generally to non-proximate causation.

The second problem with the causal realist strategy of defining events on continuous intervals is, I believe, instructive. The upshot of this problem for the realist is not to reject the strategy outright, but rather to downplay the emphasis on events in a realist account of causation. Once one sees that on the realist's second response to incoherence-type arguments there are no such things as proximate causes and effects, the demarcation of specific events that are the relata of causation becomes a fairly arbitrary matter. One may define A on $[t_1, t_2)$ and B on $[t_2, t_3)$ if one wishes, but one could just as easily choose different time intervals without incorrectly describing the causal facts of the matter. In other words, there is no constraint here based on correctly describing some mind-independent entities, A, B, and their causal relation. One might just as well choose to identify A with a different time interval and do the same for B. Any choice will do so long as A is defined on an interval that is open up to the instant at which B definitively begins.

These considerations serve to undermine the idea that events *qua* ontological category of entity are the fundamental relata of causal relations. Here again, the idea that there is a special ontological category of entities, events, that stand in a special, causal relation, should begin to ring hollow in realist ears. For these events can be sliced up in almost any way one pleases. What is crucial here is the continuum along which causation occurs, not any particular temporal slices one may, for whatever pragmatic reasons, choose to recognize as events. The conventionality of the choice renders one's explanatory reliance on events a pragmatic feature of how human beings, for a variety of explanatory purposes, decide to divide up the continuum of happenings. But this is quite a distance from the idea that specific relations between particular events constitute an objective, mind-independent thing called causation.

Let us move on to the third and final response to the charge of incoherence on behalf of traditional versions of causal realism, which focus on relations between events. Consider once again the contiguity objection. Some causal realists will be frustrated here by the respect this discussion has shown to the challenge the contiguity objection provides. After all, they will say, the objection simply misses the point! It is irrelevant whether *A* and *B* are strictly contiguous. Earlier I suggested that contiguity is a requirement if one is to think of the connection between *A* and *B* in terms of necessity (in the sense of sufficiency), for, lacking contiguity, one cannot preclude the possibility of interventions between them. However, the question of whether something could intervene so as to prevent *B*, some causal realists will claim, is a red herring, because in the circumstances there *is* no intervention. In the circumstances, *A* is sufficient (and/or necessary, depending on how *A* is characterized precisely) to bring about *B*. In any given case, or so this claim goes, by '*A*' one means to refer to a collection of factors that is relevant to bringing about *B*, and which, as it happens, excludes factors that would prevent *B*.

Two things should be noted here. Firstly, by appealing to what is causally necessary 'in the circumstances', this last response to the contiguity objection appears to conceive of causation in a rather singular manner. That is, it describes specific instances of causation in which potential interventions are ruled out by fiat. Since this response gives so much weight to the presence of specific conditions and the absence of intervening ones in particular cases, it is unclear how the resulting picture of causation might be generalized in order to yield an account of general causation, in terms of laws. It is unclear, for example, what the identity criteria should be for the types of events described by such laws, because causation is here described in terms of the *absence* of potential interveners, whatever they may be. The same observation applies to the second causal realist response to the charge of incoherence, in which events are defined on continuous temporal intervals. In this case also, necessity will have to be explained in terms of specific circumstances that obtain and others that do not. Both of these responses focus on the conditions preceding *B* in specific cases, and thus describe very particular sets of circumstances.

A final point about the last events-based answer to the contiguity objection is closely related to the one just made. What this response does, in fact, is to relocate the idea of causal necessity from the sphere of relations between events, *per se*, to that of various circumstances that make up particular events. In making this move the causal realist effectively downplays the traditional emphasis on events, and now turns a spotlight on to specific conditions that obtain or are absent. In other words, in order to give a promising response to the charge of incoherence, the causal realist has shifted from talking about relations between events to talking about specific combinations of *properties*. This insight forms the basis of a different and better proposal for causal realism. As we shall see, on this change in emphasis from relations between events to a finer-grained inquiry in terms of relations between properties, various options for an account of general causation emerge straightforwardly from the context of singular causation. And as it turns out, this change is grist to the mill of semirealism.

4.4 A BETTER ANSWER: CAUSAL PROCESSES

If arguments for the incoherence of causal realism seem compelling initially, it is only because of a lack of refinement in the traditional realist characterization of causation as a relation between events. Though misleading, however, the traditional picture is not in the wrong ballpark entirely. The problem with it is that it privileges the role of events in giving an account of causation, and this pays insufficient attention to the precise metaphysical details. Focusing on events has the unfortunate consequence of obscuring the role played by those properties of things one takes to explain their behaviours. These properties are no strangers in the context of this book, of course. They made their first appearance in Chapter 2, in my consideration of the evolution of scientific realism. Semirealism, recall, is first and foremost a realism about causal properties and their relations. So it is no surprise, perhaps, that an investigation into the nature of causation, one of the key supports on which scientific realism rests, should turn them up once again. Once one appreciates the role of causal properties in a realist account of causation, I will argue, the kinds of worries that give rise to incoherence objections simply do not arise in the first place. This understanding will provide a framework for causal realism, and vindicate it against the charge of incoherence.

When I claim that incoherence-type arguments are premised on a realist account that pays insufficient attention to certain details of causation, I have in mind primarily the objections of contiguity and regress. By spelling out the details and rising above these objections, however, I will also uncover a partial response to the demand for a causal mechanism. Let us turn to these details now. Descriptions of causal phenomena in terms of relations between events are useful for many purposes, but it is not events *qua* events that "do the work" of causation. Events commonly feature in descriptions of causation *because* they incorporate causal properties of objects. Referring to events as the relata of causation makes sense of much of our phenomenal experience simply because, as it happens, these things harbour the ontological ingredients, causal properties, that are ultimately responsible for causal phenomena. When one says that event A causes event B, what one is doing, in fact, is employing a coarse-grained shorthand for the details of causal interaction. This shorthand works rather well for most everyday and scientific purposes. It manages to latch on to the details in such a way as to do justice to one's coarse-grained observations of causal activity.

So what does it mean to say that causal properties 'do the work' of causation? In order to answer this question, let us recall what these sorts of properties are, exactly, according to semirealism. Unlike other putative properties such as logical, mathematical, or epiphenomenal properties, a causal property confers dispositions for behaviour. That is, a causal property is one that confers dispositions on the particulars that have it to behave in certain ways when in the presence or absence of other particulars with causal properties of their own. To summon the examples I mentioned earlier when first introducing this concept, the property of mass, for instance, confers on bodies that have it certain dispositions to be accelerated under applied forces. The property of volume on the part of a gas, for example, confers certain dispositions to vary in temperature in ways correlated with applied pressures. Causal phenomena are produced by the ways in which particulars with properties are disposed to act in concert with others, and it is this fact that realists should exploit in answering the charge that causal realism is incoherent.

In response to the objections of contiguity and regress, as I have argued, the events-based causal realist made two promising but ultimately self-undermining moves. The first was to appeal to properties of the continuum in order to avoid difficulties engendered by the temporal relations of discrete events. The second was to shift from talking about events *per se* to talking about collections of causal properties. Let us continue further in the direction of these moves and see where they lead. I believe the following applies to particulars generally, but for the sake of clarity I will focus presently on the causal interactions of objects. Here is a first pass: in causation, objects with causal properties are engaged in continuous processes of interaction. Dispositions borne by objects in virtue of their properties are continuously manifested in accordance with the presence and absence of other objects and properties. Objects with causal properties are thus in a continuous state of causal interaction, a state in which relations between causal properties obtain. For example, a volume of gas that comes into contact with a source of heat may expand in virtue of the dispositions afforded such volumes by properties such as temperature and pressure, and by doing so will come into contact with other regions of space. The property instances present in these new regions together with those of the gas will determine how both are further affected, and so on.

Now if this sort of process is what causation is all about, it is only fair that one be able to explain why realists generally take events to be the relata of causation. But we have stumbled across this explanation already. Of the continuous flux of causal activity that surrounds us, one takes notice of only certain parts, viz. parts that one finds interesting, or that are useful in the context of pursuing specific objectives, such as realizing desired states of affairs or avoiding harmful ones. And generally speaking one describes these parts in terms that are consistent with the coarseness of one's sensory appraisals of the relevant phenomena. Thus, causation is often described crudely in terms of events. One should be wary, however, of fixing ontological commitment simply on the basis of the grammatical form of these descriptions. Talk of events as the relata of the causal relation – '*A* causes *B*' – is elliptical for descriptions of aspects of continuous processes of the form '*A* precedes *B*, and the object or objects involved in *A* have dispositions, some of whose manifestations are present in *B*'. Events that are changes generally overlap multitudes of changes in the properties of the objects concerned. They occupy time slices during which objects with causal properties are engaged in continuous processes of causal interaction.

Perhaps another example will help to clarify the notion of an understanding of causation in terms of continuous processes. Take the classic case of a collision of two billiard balls. The first ball moves towards the second, which is at rest. As the balls collide the first stops and the second moves on. Here one has an event, *A*, the motion of the first ball, and an event, *B*, the motion of the second. *A* is commonly described as the cause of *B*. Given this coarse description of the facts, however, the objections of contiguity and regress apply. According to the contiguity objection, *A* and *B* cannot be temporally contiguous. For given that the motion of the first ball has an ending and the motion of the second has a beginning, and that the former precedes the latter, there are instants in time between *A* and *B*. According to the regress objection, the earlier motion of the first ball cannot be the proximate cause of the motion of the second. For one may consider the motion of the first ball in shorter and shorter durations prior to the termination of *A* without limit in search of a direct cause. Similarly, one may truncate *B* infinitely in search of a proximate effect.

Now let us increase the resolution of the description. What the contiguity objection takes to be a discrete event, the motion of the first ball, actually overlaps a continuous evolution in various properties of the ball, as some of the many dispositions it has are manifested in light of other property instances with which the ball "comes into contact" – properties of the table, surrounding air molecules, and so on. As the first ball approaches the second, they interact in virtue of the properties they each possess, such as velocities, momenta, etc. These interactions continue in the form of a further continuous evolution in the relevant properties (and thus continuous alterations in the relevant motions) until such time as the balls are no longer within a causally efficacious range of one another. Series of discrete events are here replaced in the description by a continuous alteration of properties, each conferring dispositions for behaviour on the objects possessing them. Thus, worries about temporal contiguity between discrete, successive events are replaced by an acknowledgment of continuous processes of causal interaction. The search for events to serve as proximate cause and effect is replaced by the understanding that candidates for these things simply constitute convenient or conventional divisions of the continuum of happenings into otherwise arbitrary time slices, themselves inhabited by numerous causal interactions. A more finely tuned understanding of causation leaves aside the objections of contiguity and regress.

Furthermore, this process view is flexible enough to provide an account of causation not merely in cases of causal change, but also in cases involving static states of affairs that some think should be diagnosed as causal also. Here descriptions of causal stasis can be given in terms of equilibrium relations, manifested in accordance with the dispositions conferred by the relevant properties. The process account might even be extended to encompass the more controversial idea of simultaneous causation, where one state or event appears to cause another, co-temporal one, or where two or more co-temporal aspects of the same state cause one another. As an aside, however, though the proposed view of causation allows for such explanations, I suspect that most if not all apparent cases of simultaneous causation (such as Kant's example of placing a lead ball on a pillow and thereby, simultaneously, causing an indentation) can be redescribed in non-simultaneous terms. For example, one might wish to causally explain why a gas occupies a specific, constant volume by noting that it has, co-temporally, a specific temperature and pressure. But the static volume of the gas could be explained instead by appealing to its prior temperature and pressure and the dispositions these confer. One could just as well explain why the gas has a certain temperature (pressure) by appealing to the dispositions conferred by its prior volume and pressure (temperature). Here various causal properties of the gas are stable, in keeping with the dispositions they confer on the gas itself.

Still, worries about the contiguity and regress objections may persist. One might wonder why a sceptic could not urge these same arguments again, but this time with continuous processes of interaction as their target. Recall, however, that crucially, the contiguity objection exploits the premise that causal relata are *discrete*. By adopting the idea that causation is a process, the causal realist is trading in an account in terms of successions of discrete events for one in terms of continuous processes. "Trading in" is a more radical suggestion than perhaps it has appeared thus far. To embrace causation as a continuous process is to view processes as causally fundamental or basic. Making the switch is not a matter of simply replacing large-scale events in descriptions of causation with micro-events made up of changes in the properties of objects. To understand causation as a process is to *preclude* the description of causation in terms of relations between discrete events, except as elliptical for descriptions of aspects of processes. That is why the contiguity and regress objections do not arise here. Of course, this does not prevent the causal realist from carving events out of parts of causal processes and calling them causes and effects, for the sake of convenience. People do this very effectively, as their success in completing everyday and scientific tasks confirms. The reason that speaking of causation as a relation between events works, however, to the extent that it does, is that events overlap continuous interactions that together compose processes.

On the process view, the causal realist is thus immune to the objections of contiguity and regress. What about the demand for a causal mechanism? The sceptic is asking here for a more detailed account of what precisely is taking place when one thing is thought to give rise to another causally. This account should include a description of some sort of mechanism for the link between causes and effects, to which the supposed *de re* necessity of causation is "attached", or from which it emerges in some way. This demand for a causal mechanism, I suggest, is partially addressed by the metaphysics of causal properties. With respect to mechanisms, the most a causal realist (or anyone for that matter) can say is that causally efficacious events incorporate objects with property-conferred dispositions, and the occurrence of subsequent effects can thus be understood in terms of manifestations of the relevant dispositions of the objects involved. The process view, while de-emphasizing the role of successive events, should not of course be construed as uninterested in the fact that some alterations follow others in time. On the contrary, what one is often most interested in are the ways in which the states of objects evolve. States change, but explaining precisely *how* such change occurs is something that one can only say so much about.

This is to concede Hume's point that ultimately one has nothing like a "picture" of what is happening when one thing brings about another beyond that which is observable or, one might add, detectable. That is why the demand for a causal mechanism cannot be fully satisfied. Objects with causal properties are disposed to behave in certain ways when in the presence or absence of other objects with properties of their own. Causal phenomena are produced by the ways in which property-conferred dispositions are linked to one another, and noting this may be the best causal realists can do.[3] So far as they are concerned, however, this best is rather good, and as we shall see, saying just this much opens the door to several accounts of *de re* necessity. The empiricist sceptic, on the other hand, disapproves not only of the prospect of necessity in the world, but also of the unqualified realist appeal to dispositions. Such talk is acceptable to empiricists only under the qualification that dispositional language is used merely as a shorthand for conditionals regarding what happens in specified circumstances, and certainly not as an invocation of causal powers. Exercising these sorts of epistemic principles is part of what it is to be an empiricist. Realists, conversely, favour different epistemic principles, and these do not require that the demand for a causal mechanism be answered to the empiricist's satisfaction. Let me clarify this further.

Although different with respect to their proximity to empirical investigation, inferences concerning unobservables in the scientific context may provide an interesting analogy for some more deeply metaphysical inferences in the context of scientific realism. In both cases, many of the unobservables whose existence realists routinely infer on the basis of their explanatory virtues are grasped metaphorically. When one asks for a description of an electron, not in terms of what one interprets as empirical measures of their quantifiable properties, but rather in terms of a deeper, qualitative picture, one is told to think of them loosely using several different, illustrative metaphors: particles, waves, clouds, etc. The metaphors one employs to fill out one's conceptual pictures of how things give rise to other things causally are also vague, and this is an interesting epistemic fact, but not one that by itself, for the realist, has any definite ontological implications. Realists do not exclude unobservables that play important explanatory roles on the grounds that the metaphors they use to conceptualize them are less precise, descriptively, than an empiricist might otherwise demand.

---

[3] Importantly, there are other questions regarding causal processes about which one can say much more. How does one identify processes in nature? How are they modelled in the sciences? How are they manipulated so as to give rise to desired outcomes? How do they furnish explanations? For work on these and related issues, see Pearl 2000, Spirtes, Glymour, and Scheines 2001, and Woodward 2003.

An electron is no ordinary particle, wave, or cloud. It is a particular whose causal properties stand in certain relations, and one's knowledge of these relations allows one to explain the electron-related phenomena one detects and creates in scientific contexts. An analogous point can be made by causal realists regarding dispositions and causal necessity. Here too one employs metaphors. What is the notion of a power, after all, if not an attempt to render intelligible, metaphorically, the notion of causal necessitation? The causal realist maintains that the explanatory benefit of causal realism is crucial in meeting the demand for an explanation of why non-accidental regularities occur. It is an empiricist's prerogative to complain that one has no precise "picture" of causal "bringing about", but for the realist, to seek precision here is to look for it in the wrong place. That is not to say that the tenability of these metaphysical explanations can be taken for granted, and the analogy to scientific unobservables breaks down if pressed. It would be peculiar, for example – a category mistake – to think of causal necessity as a particular, as one does an electron. One detects and measures the properties of the latter, but not of the former, and such procedures significantly bolster epistemic commitment in scientific contexts. How then is this *de re* feature of causal processes to be understood? Likewise, a realist about dispositions views them as ontologically on a par with other properties, but whether this understanding is defensible is a matter requiring further attention.

Several proposals have emerged concerning objective causal necessity. The most widely discussed of these make up a family of views that have in common the idea that causal laws (and laws more generally) are relations between properties. This approach is commonly associated with Fred Dretske (1977), Michael Tooley (1977, 1987), and David Armstrong (1983). Although these authors vary in their ontological commitments and in other details, each of them gives an account of a form of necessity that characterizes certain relations between properties. Another approach is taken by Sidney Shoemaker (1980, 1998) and Chris Swoyer (1982), who argue that what make a causal property the property that it is are the dispositions for behaviour it confers on the objects that have it. On this view, causal necessity follows from the nature of property identity. Objects with the same causal properties are disposed to behave in the same specific ways, simply in virtue of having those properties; objects that behave differently in exactly similar circumstances must have different causal properties. On both of these approaches, instances of singular causation are closely linked to general or law-like causation, since properties and relations instantiated in a given instance may also be present at other times and places and involve different particulars, thus generating regularities.

These views are controversial and require careful consideration. I began with the claim that the very idea of causal realism is incoherent, and what I hope to have shown thus far is that, by understanding causation in terms of continuous processes, the realist is safe from the incoherence arguments that plague events-based accounts of causation. By relocating necessity from relations between discrete, successive events to the relations of causal properties, the realist about causation takes one step forward. The questions of what further position best describes objective causal necessity and whether scientific realists must be realists about dispositions, or may instead think of them in deflationary terms, are tackled in Chapter 5. The burden of the present discussion has been to show that these further matters can and should be explored, free of the incoherence objections whose charges I aimed to dispel. Before moving on to consider dispositions and laws, however, let us undertake one last exercise in order to clarify the idea of a process account of causation. The view I have sketched suits the causal realist, but as process theories go, this one is not the only game in town. Process theories may also be advocated in the service of empiricism, and by considering the differences between these accounts one gains yet more insight into the differences between causal realists and their empiricist critics.

4.5 PROCESSES FOR EMPIRICISTS

In giving an account of causal realism, I have described processes as systems of continuously manifesting relations between objects with causal properties and concomitant dispositions. This is to use 'causal process' as a term of art, however, for there are several established uses of the term. Wesley Salmon also thinks that the traditional notion of causation as a relation between events is 'profoundly mistaken' (1984, p. 138), and he and Phil Dowe (2000) both offer process theory alternatives. Their concepts of a process, however, are different from and more specific (arguably most relevant to causation in physics) than the one I have described here. In concluding this chapter, I would like to clarify the different questions to which these various process theories are addressed. Salmon and Dowe give somewhat diverging accounts, but since the differences are immaterial to my present goal in discussing them, I will speak of their views together here, as describing Salmon-Dowe (SD) processes. Though Salmon identifies himself generally as a realist, I believe the SD account is tailor made for an empiricist approach to causation, and so it should come as no surprise that it differs from the causal realist position as I have described it. Nevertheless, I will suggest that the realist's description of processes happily incorporates SD processes as part of a broader-ranging approach to causation.

The SD replacement for an account of causation in terms of events is based on two fundamental causal concepts: process, and interaction. Salmon's concept of a process is inspired by Russell's (1948, p. 477) notion of a 'causal line', involving the persistence of an object: 'Throughout a given causal line, there may be constancy of quality, constancy of structure, or a gradual change of either, but not sudden changes of any considerable magnitude.' This is vague, but Salmon goes on to describe processes more precisely. In his most considered view, shaped in large measure by exchanges with Dowe, a process is something that carries or transmits a conserved quantity such as mass-energy, linear momentum, angular momentum, or electric charge.[4] Causal processes propagate causal influences. They are altered in causal interactions, where conserved quantities are exchanged between processes. Interactions occur only when causal processes intersect one another spatio-temporally, and though such intersections are necessary, they are not sufficient conditions for causal interaction. Interactions are the means by which modifications to causal processes are realized, and processes are the means by which the conserved quantities whose exchange constitutes an interaction are transported from one space-time location to another.

Drawing from this brief introduction, let me summarize the SD account in terms of the following definitions:

SD1   A causal process is a world line of an object that transmits (carries) or possesses (instantiates) a non-zero amount of a conserved quantity at each space-time point of its history.[5]

SD2   A causal interaction is an intersection of world lines at which a conserved quantity is exchanged.

---

[4] Salmon first described causal processes as ones that are capable of transmitting a mark, which is a signal, information, or a modification in structure (summarized in Salmon 1998, pp. 193–9). He abandoned the mark criterion in response to Dowe 1992, which introduced the conserved quantity view. Salmon's later position that processes transmit invariant quantities appears in Salmon 1998/1994, and his acceptance of conserved quantities in Salmon 1997.

[5] To be precise about the differences in their views here, Salmon holds that objects must transmit, and Dowe that they need only possess, a conserved quantity. Only Salmon gives the explicit qualification of a 'non-zero amount'. Arguably these are not substantive differences, but I will not consider the question here.

While the SD account views causation as a mind-independent phenomenon, it is silent on the issue of *de re* necessity. In this respect it is Humean. As I have noted, the intersection of two or more SD processes is a necessary but not a sufficient condition for causal interaction. Some processes pass through one another without interacting at all. Why then, one might ask, do some intersections yield interactions while others do not? In true Humean spirit, one cannot say; there are merely regularities. Conversely, the causal realist hopes to explain why specific interactions take place on some occasions and not on others. On the proposal I have outlined, the realist has a framework for such explanations, for on this view something can be said about the sufficiency conditions for causation. Specific interactions take place whenever circumstances favour the manifestation of the relevant dispositions conferred by the causal properties of objects. Causal interactions simply *are* dispositions being manifested. In Humean fashion, though, Salmon disavows causal powers. Causal processes, he says (1984, pp. 202–3), simply 'carry' probability distributions for different interactions, and he calls these probabilities 'propensities'. There is a perfectly Humean reading of probability in terms of frequency, of course. The idea is that probabilities can be understood simply in terms of the frequencies of measurement outcomes. But the term 'propensity' does rather suggest a power or disposition – precisely what Salmon denies.

Both Salmon and Dowe hold that their accounts shed light on the connection between causes and effects. As Salmon (1984, p. 155) puts it, 'The propagation of causal influence by means of causal processes *constitutes*, I believe, the mysterious connection between cause and effect which Hume sought.' SD processes, however, do not constitute Humean causal connections. Consider the following sequence of events: one strikes the white ball with a cue, the white ball moves in the direction of the black ball, the white collides with the black, the black moves on. Let us idealize the example so that the motion of the white ball strictly satisfies SD1. That is, let us assume it exchanges no linear momentum with the table or surrounding air molecules *en route* to the black ball, thus constituting an SD process between the interactions with the cue and the black. It is immediately apparent that this is not what Hume has in mind when he speaks of causal connections. For Hume, the striking of the white ball, the subsequent motion of the white, the collision with the black ball, and the subsequent motion of the black are all events. Hume sought necessary connections *between* these events, and found none. By simply relabelling some of these events 'processes', as described in SD1, and calling them causal connections, Salmon misrepresents the very nature of the connection Hume dismissed.

Dowe also maintains that his process view answers the question of what connects causes and effects, though he does not claim to be answering Hume *per se*. He (2000, p. 171) defines causal connection as follows: 'Interactions I_1, I_2 are linked by a causal connection by virtue of a causal process p only if some conserved quantity exchanged in I_2 is also exchanged in I_1, and possessed by p.' Causes and effects are understood here as events or facts that 'involve the possession, or change in value of, some conserved quantity'. In this last quotation Dowe appears to identify causes and effects with both processes, described in SD1 in terms of the possession of conserved quantities, and interactions, described in SD2 in terms of changes in the values of such quantities. However one understands causes and effects here, though, the moral is the same as that derived a moment ago in the context of Salmon's position. The SD view does not furnish an account of what Hume would call necessary connection, nor does it offer a framework for explanation with which to facilitate an account of the sort of *de re* necessity causal realists are after. The SD view is a process theory well suited to empiricists, and as such, the motivations of causal realism are foreign to it.

That is not to say, however, that causal realists should take no interest in the idea of SD processes – on the contrary! In a large range of cases, particularly in physics, the kinds of properties most relevant to the processes I described in connection with causal realism are ones that concern the possession and exchange of conserved quantities. Having determinate values of energy, momentum, charge, etc., confers dispositions on the objects that have these properties, and the continuous manifestation of these dispositions in concert with the presence and absence of other objects and their properties is what constitutes realist causal processes. To put it simply, a description of SD processes gives a partial account of what is happening in these cases, and causal realists attempt to explain further how these things happen in the first place, and why they happen at all.

This subsumption of the SD account will typify, I expect, the causal realist's attitude towards any theory that aims to explicate causation exhaustively in terms of non-causal categories of things. Douglas Ehring (1997), for example, describes causation in terms of the activities of tropes (property instances): their persistence, destruction, fission, fusion, transference, and exchange. But trope theory is an account of properties, and so trope causation is further grist to the mill belonging to the causal realist, and thus potentially to the semirealist. It would be wrong to close with the impression of a wholly irenic state of affairs, however, for the subsumption by causal realism of descriptions of causation compatible with an empiricist approach is, of course, asymmetrical. Realists may absorb empiricist descriptions, absent the view that there is nothing more to say. But empiricists will always oppose the primitive causal concept of a power, or disposition, which seems so central to the proposal for causal realism given here, in response to the arguments for the incoherence of causal realism with which we began.

CHAPTER 5

Dispositions, property identity, and laws of nature

5.1 THE CAUSAL PROPERTY IDENTITY THESIS

Causal properties are the fulcrum of semirealism. Their relations compose the concrete structures that are the primary subject matters of a tenable scientific realism. They regularly cohere to form interesting units, and these groupings make up the particulars investigated by the sciences and described by scientific theories. The continuous manifestations of the dispositions they confer constitute the causal processes to which empirical investigations become connected, so as to produce knowledge of the things they study. Scientific realists reach beyond the observable to claim knowledge of certain unobservable properties, structures, and particulars, and by doing so enter the speculative waters of metaphysics. Unlike some metaphysical commitments, being a realist does not require a wetsuit or an oxygen tank, but as I have suggested before, it is not surprising that many realists are not content merely to wade in the shallows. The portrait of realism sketched in Part I places a great deal of weight on certain meta-physical supports, and I have set out to demonstrate the internal coherence of the position by elaborating a unified account of these underpinnings, pointing out along the way certain comparisons to other possibilities with respect to capacities for explanation. Earlier I suggested that the most important aspects of the metaphysical foundations of realism are the idea of causation and the idea of natural categories of things, or kinds. I began a consideration of these issues by examining the nature of causation. We are now in a position to extend this discussion and to connect it to the topic of kinds.

Further motivation for this extension comes from the crucial distinction between abstract and concrete structures, which arose in the context of structural realism. Abstract structures, recall, are formal properties of relations, and thus higher-order properties of particulars; one and the same abstract structure may be instantiated by many different concrete structures. A concrete structure, on the other hand, consists of a relation between first-order properties of particulars, and when first-order properties are related in certain ways the particulars having these properties are generally thereby also related in some way. I say 'generally' simply in recognition of possible counter-examples – tolerance is more laudable than intolerance; this does not entail that every tolerant person is more laudable than every intolerant person – but I believe the point stands for cases of interest here. If the volume of one chocolate fudge brownie is greater than the volume of another, for example, the former brownie is larger than the latter. (The former is also more desirable, but unlike volume this is not an intrinsic property of snacks.) Semirealism advocates a knowledge of concrete structures, and it is part of the concept of these structures that in order for two of them to be identical, they must not only have the same formal properties, but must also concern the same *kinds* of relations and relata. In order to unpack their realism about concrete structures and various particulars, realists thus require some understanding of what it means for instances of causal properties to belong to the same kind, and likewise for things such as objects and events. In this chapter I will take up the first of these challenges, and reserve the second for Chapter 6.

Let us begin by remembering that realists are interested in causal properties as opposed to putative, non-causally efficacious ones. This excludes possibilities such as logical, mathematical, and epiphenomenal properties from the present discussion. In the context of scientific knowledge the realist is interested in properties that confer dispositions for behaviour. Particulars with causal properties such as volumes, masses, charges, velocities, and so on are thereby disposed to behave in certain ways in the presence and absence of other particulars and properties, and it is these properties and resulting causal processes that scientific theories describe. What is it, then, that makes a causal property the property that it is?

There are two general approaches to the question of property identity, and before turning to the one I prefer, let me mention briefly the one I will not consider in detail. Armstrong (1999, pp. 26–7) calls this view the 'categoricalist' theory of properties, and characterizes it in the following way: 'Natural properties have a nature of their own, and it is at least metaphysically possible ... that the same properties are associated with different causes and effects [and] that different properties are associated with the very same causes and effects.' The most obvious reason for unease about the categoricalist approach is the idea that properties have fundamentally mysterious 'natures of their own', in terms of which property identity is to be understood. Whether in the actual world or (in the jargon) across possible worlds, if the identity of a property is independent of the behaviours of particulars having it, what makes that property the property that it is has nothing to do with its role in causal processes. On the categoricalist view, the identities of properties are determined by un-analysable natures about which nothing substantive can be said. The idea of a primitive principle of property identity or 'quiddity' is analogous to the idea of haecceity or primitive this-ness which I mentioned earlier as one possible account of the identity of objects. Many philosophers have argued against this abstruse implication of the categoricalist theory of properties, and I will not consider it further here.[1]

The second approach to property identity requires no such appeal to unknowable natures and is therefore, I believe, more plausible *prima facie*. As it turns out, this second approach also describes causal properties in terms that fit especially neatly with the commitments of many realists. Two of its most compelling features are that both a role for dispositions and an account of *de re* necessity, which surfaced in Chapter 4 as possible desiderata for the realist, emerge simply and without further metaphysical commitment merely from an account of what it is to be a causal property. If tenable, this view would serve as a metaphysically minimal yet powerful conceptual tool. Despite its apparent promise, however, the approach is by no means free of alleged worries of its own. This chapter is dedicated to elaborating and dissolving these concerns, and by so doing, further charting my proposal for the foundations of realism. Let us turn to these tasks now.

The account of properties I will defend stems from path-breaking work by Sydney Shoemaker, in which he argues that the necessity of causal relations follows from a specific understanding of the nature of causal properties. Once this view is stated, it is immediately clear that the necessity it affords is generalizable. That is, the necessary character of causal phenomena in the context of singular causation immediately yields an account of general causation – an account in terms of laws of nature. This fits nicely with the motivation of causal realists for thinking that there

---

[1] For example, see Shoemaker 1980, Black 2000, and Bird 2005. There is also the possibility of what Armstrong calls a 'double-aspect' theory, due to C. B. Martin, discussed in Armstrong, Martin, and Place 1996. Here, property identity depends on both quiddities and the conferral of specific dispositions. This is a hybrid of the categoricalist theory I am leaving aside and the dispositional view I will defend. Since the double-aspect theory also adopts hidden, inner natures, it too suffers in my estimation by comparison to the dispositional view, versions of which are also discussed by Swoyer 1982, Ellis and Lierse 1994, Elder 1994, Mumford 1995, and Ellis 2000.

is such a thing as causal necessity in the first place. For the causal realist, the existence of certain patterns in nature calls for an explanation. The explanation I have presaged thus far is that particulars have properties in virtue of which they participate in causal processes, and these same properties can be present at other times, places, and in different particulars, thus generating regularities. Every case of singular causation is thus an instance of general causation, because singular causal processes incorporate instances of properties that are subsumable under general laws. Understood this way, as we shall see, the second approach to property identity yields yet another benefit: a straightforwardly emerging account of laws.

So what is this understanding of the nature of causal properties that immediately yields conclusions about laws of nature? Here is a rough opening sketch. To say that a particular has a certain causal property is to say that it is disposed to behave in certain ways in certain circumstances, and that all particulars having this same property are likewise so disposed. By circumstances I mean the presence and absence of other causal properties, both of the particular in question and of other particulars. Some of the processes elicited by these circumstances are experienced by us in the form of detected regularities. These regularities unfold in accordance with systems of laws which one attempts to describe using linguistic expressions, often in the form of mathematical formulae. Causal laws are relations between causal properties. As will become evident, this account of laws has certain similarities to the family of views pressed by Dretske, Tooley, and Armstrong, but with important differences which I will consider in sections 5.4 and 5.5. For the sake of simplicity, and in keeping with the convention established in previous chapters, I will use objects as my paradigm example of particulars henceforth.

That was an opening sketch. In giving it I have already incorporated three crucial assumptions, and it will serve us to make them more explicit. First, there is an assumption here about laws. They are not sentences, statements, or linguistic entities of any kind. They are aspects of nature that *make* the linguistic devices one employs to describe them true or false. This is an ontological as opposed to a linguistic conception of laws. Those who adopt a linguistic conception use the term 'law' for what are in fact descriptions of laws, according to the ontological conception. Adopting the ontological view, I will refer to such descriptions as law *statements* rather than laws, in order to maintain a clear distinction between descriptions of laws and laws themselves. Admittedly, for many purposes there is little harm in conflating these terms, but it is important to appreciate the distinction, and not merely for the sake of being pedantic! As it turns out, the distinction is important for metaphysical reasons which will become clear later in connection with the idea of vacuous laws.

A second assumption here concerns the nature of causal properties, and constitutes the heart of this alternative to the categoricalist approach to property identity. Contained in my rough opening sketch is not only the notion that causal laws comprise relations between causal properties, but also the notion that knowing such laws allows one to distinguish and identify properties as well. A causal property can be identified as the property that it is in virtue of its relations to other properties. The conjunction of all causal laws thus specifies the natures of all causal properties. Perhaps the clearest statement in the general spirit of this position is given by Shoemaker (1980, p. 133), who says that 'the identity of a [causal] property is completely determined by its potential for contributing to the causal powers of the things that have it'. Where Shoemaker speaks of causal powers I will continue to talk about dispositions, but there is no substantive difference. Let us refer to this understanding of property identity as the dispositional identity thesis (DIT).

DIT is the general idea I hope to defend, but before attempting to refine my opening sketch of this approach, a third and last assumption requires attention. I have liberally employed the language of dispositions and manifestations, but as I indicated earlier there are different ways of interpreting such language, and it is time now to clarify this situation. It is not entirely clear whether Shoemaker is a realist about dispositions, and it is further unclear whether a scientific realist need be. Indeed, it is slightly unclear what it means to be a realist about dispositions in the first place. The primary reason for this fundamental lack of clarity is that the concept of a disposition is commonly described by means of two different contrasts, and these modes of description are not equivalent. A brief consideration of them may help to clarify the distinction between realism and antirealism about dispositions.

Dispositional properties are often contrasted with so-called 'categorical' properties. The difference is usually explicated in terms of the manner in which they are described: the former in terms of what happens to objects under certain conditions, and the latter without reference to any happenings or conditions. Canonical examples of dispositions are properties like fragility and solubility, described respectively in terms of what happens to certain objects when treated roughly, and what happens to certain things when placed in solvents. Categorical properties, on the other hand, are described in terms of static features of particulars such as their dimensions (length, volume), shapes (rectangular, cylindrical), configurations or arrangements (molecular structures), and so on. By itself, this distinction between the dispositional and the categorical has no implications for the issue of realism. Some hold that if there are such properties as dispositions they must be "grounded" in categorical properties, as some suggest solubility is grounded in the molecular structures of soluble compounds. Others insist on the possibility of "bare" dispositions which require no grounding. Some think the dispositional–categorical distinction does not apply to properties at all, but merely to the predicates that name them, and that all causal properties are describable either dispositionally or categorically in principle if not in practice. As far as semirealism is concerned, these are open questions, possible answers to which are all consistent with DIT.

A second way in which dispositional properties are often characterized is by means of a contrast with so-called 'occurrent' properties, and this distinction *does* have implications for the choice between realism and anti-realism about dispositions. An occurrent property is one that genuinely exists. Its instances are part of the furniture of the world. Those who accept the contrast between dispositional and occurrent properties thereby adopt some form of antirealism about dispositions. This is often the favoured position of empiricists, who generally have nothing against the use of dispositional language so long as one is not misled by such figures of speech into thinking that dispositions are real. On this view, saying that an object has a disposition is simply elliptical for saying something about how it would behave under certain conditions. It is in this tradition, for example, that Gilbert Ryle (1949) famously describes dispositional ascriptions as 'inference tickets', because they facilitate inferences concerning what happens to objects in specified circumstances. For the empiricist sceptic, the semantics of dispositional language is given purely in terms of manifested behaviours.

Scientific realists may join empiricists in opting for a deflationary, linguistic account of dispositions if they wish. Just as we saw in the case of objective causal necessity, however, many realists find the empiricist line unattractive here. These people instead regard dispositions as *bona fide* occurrent properties, and deny the supposition that 'dispositional' and 'occurrent' are mutually exclusive labels. Here again their reasons are guided in part by the metaphysical stance, which sanctions the pursuit of explanations in terms of the unobservable. Just as the notion of *de re* necessity serves the explanatory function of distinguishing causal regularities from merely accidental ones, the concept of dispositions helps to explain why causal processes evolve in the ways they do.

The metaphysical stance is not the only motivation for realism about dispositions, however. It is also the case that empiricist attempts to analyse them away by reducing disposition-talk to conditional sentences that exclude them are generally viewed as unsuccessful. Two well-known attempts, for example, are customarily attributed to Rudolph Carnap. The first suggests that ascribing a disposition to an object is equivalent to asserting a conditional to the effect that if certain circumstances obtain, the object behaves in a certain way. Saying that sugar is soluble is simply a way of saying, for instance, that if placed in warm water, it dissolves. But this is unsatisfactory, for if circumstances favouring the manifested behaviour are *not* present, the conditional is true (given that its antecedent is false), and one must say the object possesses the relevant disposition as a matter of logic, which is absurd. The second attempt involves a more sophisticated conditional to the effect that if certain circumstances obtain, then ascribing a disposition is equivalent to attributing the manifested behaviour to the object in question. But this does not work either, for on this view one can neither affirm nor deny a dispositional ascription unless the specified circumstances favouring the manifestation occur, and this is at odds with everyday linguistic practice in which one happily describes things as fragile or soluble even if they are in no danger of breaking or dissolving.[2]

Apart from the motivation of general empiricist sensibilities, the most famous reason for being suspicious about dispositions turns out to be no reason at all. Many worry about the reality of dispositions as a consequence of the supposed emptiness of explanations citing 'dormitive virtues', ironically dubbed by seventeenth-century French playwright Molière in *Le Malade imaginaire* ('The Imaginary Invalid'). In one scene, a great physician is applauded by the chorus when, in response to the question of why opium causes drowsiness, he responds by wisely asserting that opium has a *virtus dormitiva* – a power to cause drowsiness. Molière's physician is of course worthy of derision, but this fictive incident is often and mistakenly offered as an indictment of realism about dispositions. It is not, however. Citing a disposition can be non-explanatory in some contexts, like that of explaining why opium causes drowsiness by citing the disposition of opium

---

[2] See Mumford 1998, ch. 3, for a thorough discussion of attempts to define dispositional predicates using conditionals. For arguments in favour of realism about dispositions, see also Cartwright 1989, regarding the reality of 'capacities' in the interpretation of scientific theories, Mellor 1991/1974, and Ellis and Lierse 1994.

to cause drowsiness (!), but this tells one something interesting about contexts of explanation, not the existence of dispositions. In other contexts there is nothing empty about affirming the existence of a dispositional property to explain manifested behaviours. Consider the physician's answer in response to a different question, say that of why a patient feels drowsy after completing his or her midday routines. Here, ascribing a disposition to opium is not empty at all. It is an empirical hypothesis which may turn out to be false, and this, of course, entails that it is not empty.

Though semirealism does not require a realism about dispositions, many scientific realists are realists of both sorts, just as they are realists about some form of *de re* necessity. According to dispositional realism, dispositions are properly viewed as genuine occurrent properties regardless of whether any specific behavioural manifestations occur. As in the case of necessity, it would be wrong to insist that anyone who is inclined towards semirealism must accept dispositions also, as opposed to deflationary accounts of dispositional talk. There is room within the broad spectrum of commitment associated with scientific realism for differences on these matters. Anyone who is wary of dispositions must accept, however, that Molière's joke at the expense of his eminent physician is not a good reason for rejecting them, and that empiricist paraphrases of dispositional language do not appear to capture adequately the intentions of our speech acts. Realism about dispositions is not merely an accessory. In addition to its explanatory value, its further appeal will soon become clear in connection with an important objection to DIT. From here on, I will assume a realism about dispositions.

5.2 PROPERTY NAMING AND NECESSITY

DIT states that the identity of a causal property is determined by the dispositions it confers. In some initial moves to elaborate this view, I have attempted to redeem one of the promissory notes of Chapter 4, where I suggested that an investigation into the nature of causal properties would help to clarify the issue of what attitude the semirealist should take towards dispositions. Another promissory note remains, however, for I also claimed that reflecting on the nature of causal properties would furnish an understanding of objective causal necessity. Let us turn to this second task now. For starters, we require some terminology. Philosophical discussions of modality – that is, of necessity and possibility – make use of several distinctions in order to differentiate various modal subtleties, but for present purposes two basic distinctions are all that is required. The most fundamental one is between metaphysical necessity and contingency. That which could not be otherwise, or as it is often put, that which is the same in all possible worlds, is metaphysically necessary. The truths of deductive logic and analytic statements (ones that are true by definition), for example, are often said to be metaphysically necessary. On the other hand, that which could be otherwise or is different in some possible world is contingent.

It is within the scope of contingency that things get interesting for the realist. Recall that necessity *de re* is a feature of the world as opposed to mere necessity *de dicto*, which is a feature of things that are said or thought. Those who subscribe to *de re* necessity generally hold that the things to which it applies are ultimately contingent. (Some actually describe this necessity as metaphysical necessity, but as we shall see, what they actually intend is a qualified sort of metaphysical necessity, which even they would agree is ultimately contingent.) For the most part, those who believe that some sort of necessity is a feature of causal processes or laws of nature do not think they are metaphysically necessary, for there are possible worlds, presumably, in which causal processes and laws look very different from the sorts of things one finds and investigates scientifically in the actual world. Those who speak of *de re* necessity thus generally intend something weaker than metaphysical necessity, but this 'something weaker' should nevertheless distinguish genuinely causal processes from mere accidents or coincidences. This is a fine line to walk, and a difficult notion to articulate.

Perhaps the best way to clarify the sense of necessity sought by such a realist is to focus on the explanatory role the concept is meant to serve. This person craves a form of necessity that explains why the patterns typifying causal processes and certain relations between properties could be otherwise, but never *are* otherwise in the actual world. In a given world such as the actual one in which we live, there is *ex hypothesi* some substantive reason for the uniformity of these processes and relations. It is the presence of such a reason, in terms of which realists hope to explain the necessity they attribute to causal processes and laws, that distinguishes these things from accidents and coincidences. Since on this view causal processes and laws are ultimately contingent, it is perhaps misleading to label the fact that they are non-accidental with the term 'necessity' at all, but this usage is widespread and one is stuck with it. Capitulating to common practice, then, let me call this weaker sense of necessity 'natural necessity' (also commonly referred to as 'nomic necessity'). Natural necessity applies to that which could be otherwise, but never is in a given world for some principled reason. This reason, however, needs to be explicated if the concept of natural necessity is to play the explanatory role it is intended to fill.

As it happens, there are two established approaches to understanding natural necessity. One of these emerges straightforwardly from DIT and the other does not. I mentioned both of these accounts briefly in Chapter 4 when I suggested that there are two main approaches to the idea of *de re* necessity compatible with a causal realist view of causal processes. The first of these originates from the idea that laws generally, and causal laws in particular, are relations between properties. The authors jointly credited with formulating this account, Dretske, Tooley, and Armstrong, all hold that laws of nature are contingent. Nonetheless, each of them thinks that natural necessity is properly attributed to certain relations of properties in the actual world. Dretske (1977), for example, suggests that natural necessity can be understood simply in terms of the fact that in any given world, the natures and behaviours of particulars are constrained, because they can instantiate only properties and relations that exist in that world. On this view, the very nature of instantiation gives rise to the causal regularities one finds there. In a variation of this, Armstrong (1983) holds that natural necessity is best understood in terms of a special relation, which he labels 'N' for 'necessitation', that applies to certain relations of properties but not others.

These views are predicated on the assumption that properties are universals (abstract entities) instantiated by concrete things, and it is not obvious that this commitment is one a semirealist needs to make. Later I will suggest that deeper metaphysical questions as to the precise ontological nature of properties go beyond what is required to make sense of scientific realism. As alternatives to realism about universals, nominalists maintain that properties are best understood simply as sets of particulars, or perhaps as sets of tropes (property instances, conceived as "abstract particulars"). To the extent that different accounts are tenable here, there is no need for realists to make any one of them a component of their realism. That is not to say, of course, that all of these views are tenable. I submit, however, that this is one subject the metaphysician *qua* scientific realist can leave for others to debate, for it belongs to a depth of metaphysics that exceeds the immediate context of realism. The ontological specificity of the Dretske-Tooley-Armstrong approach thus furnishes one reason to leave it aside, for if realists can make sense of natural necessity in a way that is compatible with various accounts of the finer-grained ontological nature of properties, they thereby adopt a broader and more persuasive position. DIT, I will suggest, satisfies this recommendation.

Another reason for dispensing with this first approach applies specifically to Armstrong's variation of it. On his view, the reason certain relations between properties can be said to be subject to natural necessity is the existence of N, the special relation that obtains between these properties and not others. There is some question, however, as to whether positing a special relation of this sort actually helps to fill the explanatory role assigned to the concept of natural necessity. By itself, the existence of N does not appear to offer much of an explanation at all, and appealing to it may well amount to little more than the assertion that some relations are necessary. One might irreverently dub this worry the 'mighty biceps' objection, after an especially entertaining formulation of it by Lewis. In a discussion of the nature of necessity conferred by Armstrong's N, Lewis (1983, p. 366) offers the following diagnosis:

[Armstrong] uses 'necessitates' as a name for the lawmaking universal N; and who would be surprised to hear that if F 'necessitates' G and a has F, then a must have G? But I say that N deserves the name of 'necessitation' only if, somehow, it really can enter into the requisite necessary connections. It can't enter into them just by bearing a name, any more than one can have mighty biceps just by being called 'Armstrong'.

To put it another way, merely calling something necessary does not make it so. If the realist is to have a substantive account of natural necessity, it should consist in something more than merely giving certain features of reality a label to that effect.

DIT takes a different approach to natural necessity, and this alternative is not only open to various possibilities regarding finer-grained issues of property ontology, but also immune to anatomical, mighty-biceps-type objections. According to this view, what make a causal property the property that it is are the dispositions it confers on the objects that have it. If properties are understood this way, a simple and compelling account of natural necessity follows neatly and immediately. It is a consequence of DIT that objects with the same causal properties have the very same dispositions for behaviour, simply because of what makes these properties the properties they are. This is true of all causal properties including irreducibly probabilistic ones, such as the disposition of a radioactive atom to decay within a certain period of time with a certain probability. In the case of deterministic causal properties even stronger claims follow: objects with the same causal properties behave in the very same ways in exactly similar circumstances; objects that behave differently in such circumstances have different causal properties. They must, because property identity is determined by the dispositions for behaviour these properties confer. A moment ago I suggested that if the concept of natural necessity is to be viable, it should offer a substantive criterion with which to distinguish genuinely causal processes from mere accidents or coincidences, and this criterion should be explanatory. DIT, I believe, meets both of these expectations. The patterns that typify causal processes are not otherwise in the actual world because of the identities of the properties they incorporate. It is in the very nature of a causal property that it confers specific dispositions and not others. That is what a causal property *is*.

Relations between causal properties, which determine how causal processes evolve, must be the ways that they are because dispositions for these specific relations make these properties the properties that they are. Laws are composed of relations, the potential for which is determined by the identities of causal properties. Given that specific sets of dispositions uniquely identify each causal property, relations between these properties could not be other than they are. Thus, properties in the actual world could not have entered into laws other than the actual laws. In this way, natural necessity emerges from DIT as a consequence of its identification of specific dispositions with specific properties. The idea that there is a reason why causal regularities are as they are in the actual world is accommodated, and its explanatory power rests in a substantive claim about the identities of causal properties.

The fact that this account of necessity is stronger than the natural necessity described on the general approach of Dretske, Tooley, and Armstrong leads its advocates to call it 'metaphysical' necessity. On the Dretske-Tooley-Armstrong view, one and the same causal property can stand in different relations to other causal properties in different possible worlds, since their identities are determined by something other than the dispositions for relations they confer, and this possibility is ruled out by DIT. It is for this reason that advocates of DIT-type approaches suggest that causal laws are metaphysically necessary, but this must be understood in a carefully qualified way. *Given* that a possible world is inhabited by the same causal properties as those populating the actual world, causal laws there will be the same as they are here. In other words, causal laws are the same in all possible worlds in which actual-world causal properties are found. I prefer to call this form of necessity natural and not metaphysical, simply because it is no consequence of DIT that the causal laws of the actual world are found in all possible worlds. Nothing precludes the possibility that laws might have been different, because the causal properties of the actual world, presumably, are not themselves metaphysically necessary existents. In a world inhabited by different causal properties, the relations one would there describe as laws would be, *ipso facto*, different as well. Causal laws are thus ultimately contingent here, but in a much weaker sense than is suggested by Dretske, Tooley, and Armstrong.

Nevertheless, most commentators, including Shoemaker, Swoyer (1982), Crawford Elder (1994), and Brian Ellis (2000), speak of DIT-type causal necessity as metaphysical necessity. This is most appropriate, I believe, if one adopts a linguistic as opposed to an ontological conception of laws. A statement correctly describing the relations of causal properties may be true in all possible worlds, so long as the relevant predicates continue to refer to the same properties and one does not require that such properties and relations exist at all worlds in order to make such statements true there. If one accepts these caveats, law statements may be described as metaphysically necessary. But the realist is interested in *de re*, not merely *de dicto*, necessity, and there may well be possible worlds in which causal properties and thus processes are not like those in the actual world. So, on an ontological conception of laws, one can see why it may seem appropriate to say that something weaker than metaphysical necessity applies, and that is precisely the role intended for the concept of natural necessity. (I am ignoring here the possibility that all possible causal properties and relations exist at all possible worlds. If this were the case, different and mutually exclusive sets of causal properties and laws would be co-present but causally insulated from one another. All worlds would contain indefinitely many, completely isolated worlds of causal interaction – a strange but interesting possibility indeed!)

These last clarifications of the account of necessity following from DIT are important, because one of the most common sources of uneasiness about this approach is the impression that it views laws of nature as metaphysically necessary. This conflicts with a deeply held intuition, shared by many, that there are at least some possible worlds in which laws are different from what they are in the actual world. Properly understood, however, DIT resolves this apparent tension, for on this understanding it is really only law statements that are strictly, metaphysically necessary, and even then only subject to certain conditions. Laws themselves are naturally necessary and ultimately contingent in the sense I have described. The realist is thus furnished with a helpful account of *de re* necessity. When I speak of necessity henceforth, it is this concept I will have in mind.

With these clarifications of the issue of modality in hand, we are now in a position to examine a closely related issue, regarding the ways in which properties are named. There are at least two conventions for naming causal properties that are consistent with DIT, and in the remainder of this section I will briefly consider the practical limitations the concept of necessity places on the extensions of property terms, and offer some advice regarding linguistic practice. The first of these conventions is simply to treat property terms as rigid designators. As described by Saul Kripke (1980), a rigid designator is any name, general term, or definite description that designates the same thing in all possible worlds in which that thing exists. An identity statement between rigid designators, if true, is necessarily true. On this approach, causal properties that exist in the actual world, for example, are rigidly designated both by their names and by definite descriptions in terms of the dispositions they confer here. Of course, in speaking of names and definite descriptions, I do not want to lapse into giving the mistaken impression that the necessity at issue is mere necessity *de dicto*. There is no possible world in which properties referred to by these names are related by different laws of nature, for if they were to confer anything other than the specific dispositions with which they are identified in the actual world, they would be different properties altogether.

On a second approach to naming, also consistent with DIT and its concomitant notion of necessity, causal properties are identified by their places in laws in any given world, but the existence of a set of properties does not entail any specific set of law statements across possible worlds. On first reading this must seem at odds with DIT, but the confusion is quickly remedied. Consider Mellor's (1995, p. 172) view of properties and laws, according to which 'the property M such that $F = MA$ in our world may also exist in worlds where $F \neq MA$'. There are different ways one might interpret this claim. Categoricalists about properties will interpret it in conformity with their theory, which states that property identity is determined by something other than the dispositions for relations they confer. On the other hand, one might interpret Mellor's claim in a manner consistent with DIT. In this case one would not appeal to rigid designation, but would rather view property *identity* strictly in terms of DIT, while simultaneously applying looser standards to linguistic practice in the context of naming across possible worlds. In this way, the extension of a property term like 'F' ('force') might include many strictly non-identical properties.

Perhaps the simplest way to understand this second convention is by thinking of it as a trans-world cluster approach to property classification. On this approach one accepts that a causal property is the property that it is solely in virtue of the dispositions it confers, but so long as a property confers a set of dispositions that is sufficiently similar to those conferred by its counterparts in other worlds, one refers to all of them using the same name. Properties that confer different dispositions in different worlds are here given the same name *despite* differences in the dispositions with which they are associated. They are identified as members of a class of properties on the basis of similarities in the dispositions they confer. As a classificatory strategy this seems messy to me, but one could make the case that it represents an intuitive approach to property terms, for it accommodates the intuition that properties across worlds that are sufficiently similar to one another should be classified together. It may seem natural, for example, to think of a property that figures in an inverse square law and a nearby possible world counterpart that figures in an inverse cube law in highly similar circumstances as closely related.

The choice between the first and second naming practices outlined here turns on the question of what is "properly" regarded as falling within the extensions of causal property terms. But this is merely a matter of convention, and not by itself consequential as regards DIT or necessity. In philosophical discussions of properties and laws, some use 'mass' to refer only to properties that confer the same dispositions as the ones mass confers in the actual world, while others use the term to pick out properties across worlds whose associated dispositions are sufficiently similar. The former view, I think, is clearly more attractive. On the latter, trans-world cluster scheme, awkward questions are invited regarding where to draw the line between properties that confer similar enough sets of dispositions (thereby figuring in similar enough laws of nature) to count as part of the extension of a property term, and those that do not. On the former view, however, if *per impossibile* the property one calls 'mass' were associated with different dispositions in some possible world, it would clearly *not* be what *one* calls 'mass'. Awkward questions about the reasonableness of criteria for determining the extensions of property terms are avoided. In the following discussion, I will adopt the former, simpler approach.

A preliminary sketch of DIT is now complete, and I have considered both the place of dispositions and the idea of necessity in support of semirealism. A realist's work is never done, however. Having sketched an outline of what it is to be a causal property, there are several challenges to face, for criticisms of DIT remain to be answered and the details of its connections to the idea of laws of nature stand in need of development. Let us confront these challenges head-on, first by examining arguments that aim to expose DIT as an untenable account of properties.

5.3 OBJECTIONS: EPISTEMIC AND METAPHYSICAL

DIT asserts that the identity of a causal property is wholly determined by certain dispositions for relations with other properties, or in other words, by the dispositions it confers for behaviour on the things that have it. Several important concerns, however, suggest that this view of properties cannot be maintained, and these objections fall into broadly epistemic and metaphysical camps. Let me begin with the epistemic. A first challenge is addressed to the question of whether it is possible to know properties on this view. Consider for example Rosenberg's (1984, p. 82) contention that DIT holds a knowledge of causal properties hostage to a knowledge of the causal laws in which they figure. A generalized version of the argument goes this way. If the identity of a property is determined by the relations of which it is capable, as described by law statements, then citing these law statements is a necessary precondition for picking out or identifying properties. However, our past, present, and foreseeable future stock of knowledge contains nothing like a complete specification of causal law statements. Furthermore, it seems that we have been, are, and presumably will continue to be able to identify causal properties. Given that one has a knowledge of these properties despite impressive gaps in one's knowledge of laws, DIT must be mistaken.

This argument is premised on a misconception. It is easy to appreciate its allure, however, because one of its premises is highly intuitive. The proponent of the argument is correct to maintain that one is often successful in identifying causal properties without knowing all of the laws in which they figure. For example, one can know an object to be red despite not being able to describe the causal processes by which the light reflected from it interacts with one's visual system. Similarly, regarding less controversial instances of intrinsic properties such as charge and volume, causal properties can be identified prior to a complete specification of laws. One can even know that two properties are the same in the absence of a detailed knowledge of their relations, in just the way that samples of the same colour can be matched together simply by looking at them. The premise that one is capable of identifying causal properties in the absence of a complete knowledge of laws is certainly compelling.

Imagine that two laboratory samples share a property that one can detect only via microscopy. One may determine that these samples share one and the same property without knowing all of the relations of which that property is capable. Admittedly, judgments of property identity may well require the assumption that causal laws are in effect. For example, the judgment that two samples viewed microscopically share a causal property assumes the causal functioning of the microscope. A detailed knowledge of the relevant laws, however, having to do with causal processes involving the samples and the microscope and ultimately the microscope and our senses, is not required for this sort of judgment. Statements regarding identity and difference premised on the causal efficacy of specific properties are rarely backed up by explicit statements of causal laws, and yet causal properties are identified nonetheless. The advocate of DIT *agrees* with these observations.

The flaw in the objection resides in another one of its premises. It is a mistake to think it a consequence of DIT that a complete knowledge of laws is required to pick out or identify properties. Two points should be made here. First, in order to identify a specific causal property, it is sufficient to know that there *are* laws supporting one's inferences from detected regularities to the property in question. Detailed knowledge of the laws that describe these regular processes may come later or not at all. Take the example of detection by microscopy. It is the correct belief that there are causal laws that describe the properties whose interactions constitute causal processes linking the samples to resultant visual images, whatever these laws may be, that permits the correct judgment that two specimens share one and the same property. This is not at all inconsistent with holding that causal properties are identical in virtue of conferring identical dispositions for places in causal laws.

It is certainly the case that one cannot give an exhaustive inventory of the dispositions conferred by a property without knowing the details of all of the causal laws in which it figures. But there is no contradiction in thinking that one can identify properties without giving exhaustive inventories, and simultaneously believing that such inventories ultimately determine the identities of properties. Like anyone else, those who subscribe to DIT can measure and thus know the mass of an object, for example, without knowing all of the relations of which that property is capable, and that differentiate it (in a metaphysical, not an epistemic sense) from all other properties. A sceptical worry might arise here if it were the case that generally, different properties overlap so much in the dispositions they confer that discriminating them is impossible. One's ability to discriminate in both the sciences and everyday life, however, suggests the opposite, and even if this sceptical possibility were actual, identifying a property would generally require something much weaker than a complete specification of laws. DIT does not require that one know all laws in order to identify causal properties any more than the categoricalist theory of properties requires that one know quiddities (unknowable in principle!) in order to identify them.

A second epistemic argument against DIT, due to Richard Swinburne (1980), offers the threat of an infinite regress. I will consider a close metaphysical cousin of this concern shortly, but for the moment let us focus on the epistemic worry. It is argued that according to DIT, the attribution of properties to an object will in general require an appeal to causal powers – the dispositions properties confer – since *ex hypothesi* the identities of these properties are determined by the dispositions of objects having them to undergo certain manifestations. Thus, property attribution makes reference to the results of causal interactions, but results are properties in their own right, and to attribute these properties one must appeal to some manifestations they give rise to, and here one has the makings of a regress. One might contend that on the assumption that there is more to property identity than the dispositions for behaviour properties confer (as suggested by categoricalists), or more specifically, on the assumption that properties can be recognized independently of their causal relations, the regress could be broken. It seems impossible, however, to give any empirical content to this suggestion. If there *were* something more to causal property identity than the dispositions they confer, how would one recognize this extra something? It appears one has no other option but to ground property attributions in the causal interactions one experiences and detects. Granting this, however, is not to surrender to Swinburne's challenge, for the regress suggested does not attach to DIT.

The error of the regress argument lies in the premise that, according to DIT, in order to attribute causal properties by appeal to certain effects, properties associated with these effects must invariably be attributed by appeal to further effects. That is not the case, and is not required on the dispositionalist view of properties. Regresses of this kind are commonly short-lived, since causal chains originating with the property instances one attributes are connected, in cases where one justifiably claims knowledge of them, to one's sensory modalities. To put it another way, every case of warranted causal property attribution is facilitated by some properties that are known independently of a knowledge of their further effects. These latter property instances are the direct objects of our perceptions. Consider the everyday use of simple measurement devices. One attributes properties such as ambient temperatures and pressures by appealing to effects registered on instruments such as thermometers and barometers. The properties one associates with these effects (specific states or settings of measurement devices) constitute what one might call perceptually direct properties, since the relevant immediate effects of their instances are perceptual states on the part of the observer. Philosophers have described the contents of sense perception in many ways, but however it is best characterized, an observer's acquaintance with it does not depend on a knowledge of further effects (cf. Swoyer 1982, p. 214; Fales 1990, p. 222).

So, there is nothing inconsistent in holding that perceptually direct properties are correctly described by DIT, and yet attributed without appeal to their effects. This is not to say, of course, that there *are* no further effects – that causal processes suddenly cease. Rather, it is to say that when it comes to property attribution, the epistemic buck stops with perception. This resolution to the threat of regress brings into focus the generally overlooked but important fact that one is a participant in the causal processes that permit the attribution of causal properties to objects. One is a participant in the sense that ultimately effects must be registered *in one* for such attribution to take place, and it is here that the relevance of further manifestations runs out. The relative coarseness of one's sensory modalities distributes causal properties along a spectrum, according to the lengths of the causal chains one must exploit in order to attribute them. In cases where the property instances that interest one are instances to which one's senses provide little or no direct access, one compensates by making use of longer causal chains. That is, one employs instruments of detection. For human beings with unexceptional sensory modalities, the attribution of ambient warmth is more direct than the attribution of an ambient temperature of 27°C. In both cases, these properties are attributable even if DIT is true, and even if one's perceptual states give rise to further effects, whatever they may be.

Let us now switch tack and consider some important metaphysical challenges to the dispositional view of properties. In fact, one can think of these arguments as constituting the two horns of a dilemma. Though I did not mention it earlier, the assiduous reader may have detected an ambiguity in my opening sketch of DIT with respect to the precise nature of the relation between causal properties and dispositions. This ambiguity is present in the claim that causal properties "confer" dispositions for behaviour. There are two obvious ways in which one might explain what 'confer' means here. One might hold that dispositions are distinct from but nonetheless in some sense "attached" to causal properties. That is what some intend by the claim that causal properties are the categorical bases of the dispositions with which they are associated. On this view, causal properties can be thought of as analogous to bare particulars, though in order for this to be consistent with DIT, "bare properties" would have no quiddities (the primitive principles of property identity suggested by the categoricalist theory of properties). On the other hand, one might hold that causal properties just *are* collections of dispositions, being composed of them, as it were. To complete the analogy to particulars, one can think of this as a sort of bundle theory. The proposed dilemma then takes the following form. Given DIT, causal properties are presumably related to dispositions in one of these two ways, but neither alternative is tenable.

The first horn is shaped this way. Assume that causal properties possess or instantiate dispositions, in perhaps much the same way as they are themselves possessed or instantiated by objects. As I mentioned when first introducing the concept of a causal property, most are "many-faceted". Consider, for example, a particular value of mass, say 1000 kg. This property may figure in many different causal laws, and in each such case it does so in virtue of a different disposition conferred on the objects possessing it. To be accurate, then, one should think of causal properties generally as being many-faceted, in that more than one disposition can be associated with a given property. But the dispositions associated with causal properties, so the argument goes on this horn of the dilemma, are *themselves* properties – that is, higher-order properties – and this leads to an infinite regress. For if, according to DIT, dispositions fix the identities of causal properties, then the identities of dispositions, themselves properties, must be fixed by even higher-order properties, and so on. (The idea here that a disposition may be a higher-order property must be understood in a certain way. It does not entail that the "bare property" itself has a disposition, though this may be so in some cases according to the trope theory of properties. Rather, instances of causal properties simply "carry" dispositions of the relevant particulars.)

Closer examination, however, reveals that one is hard pressed to establish a genuine regress here. The challenge demands a response to the question: if dispositions determine the identities of their bases, what then determines the identities of dispositions? But the answer provided by the challenger in order to generate the regress, viz. that an additional order of properties is required to determine the identities of dispositions, is not a good one. The identity of a disposition is determined not by further yet higher-order levels of properties, but by the behaviours of which objects having it are thereby capable. One disposition is identical to another if and only if objects possessing them are thereby disposed to behave the same way (with certain probabilities, in the case of probabilistic dispositions) in exactly similar circumstances favourable to their manifestations. To put it another way, it is certainly reasonable to inquire about the dispositions that determine the identity of a causal property, but to seek an answer to the question of what determines the identity of a *disposition* by postulating even higher-order layers of properties is to look for answers in the wrong place entirely. The best strategy one has for identifying a disposition is to investigate causal processes involving objects one knows to have the relevant causal property. The success of this strategy is explained by the fact that the behaviours objects are capable of determine the identities of dispositions.

Let us turn now to the second horn of the dilemma. On this interpretation of DIT causal properties just *are* dispositions. That is, causal properties simply comprise dispositions of objects to act and to be affected in various ways in specific circumstances. Such properties would be composed of anything from a single disposition to a cluster of dispositions, reflecting the many-faceted nature of causal properties generally. According to DIT, dispositions for behaviour alone are what make causal properties the properties that they are. On this view, however, what are manifestations of dispositions if not further properties, which again are nothing but powers to bring about yet further causal activity? Once more, DIT is threatened with a regress. The power to bring about the set of causal properties P_1 is really just the power to bring about the power to bring about the further set of causal properties P_2, and so on *ad infinitum*. This is a metaphysical analogue to the second epistemic objection considered a moment ago (both can be traced to Swinburne 1980, and Armstrong 1983, p. 123, argues along similar lines). Note that on this horn of the dilemma, unlike the first, the various sets of properties ostensibly generating the regress are all of the same order.

Closer inspection reveals that the complaint of the second horn trades on an ambiguity, and once clarified the argument poses no threat to DIT. The ambiguity concerns the relation between dispositions and their manifestations. The clarification required is that dispositions are not *constituted* by the various developments they facilitate in causal processes. They may be identified in the epistemic sense of 'identify' by appealing to such developments, since one commonly describes dispositions in terms of their manifestations, but they should not be identified *with* such developments in the metaphysical sense of 'identity'. Dispositions are distinct from and should not be confused with their manifestations. A specific disposition may facilitate the instantiation of another, which then facilitates the instantiation of another and so on throughout a causal process, but this does not entail that the original disposition is constituted by the sequence, be it actual or merely hypothetical. Recall that realists about dispositions maintain that they are occurrent properties in their own right. The fact that a causal property may stand in relations to others, and that its identity is determined by its *potentials* for such relations, does nothing to compromise its distinct nature. Causal properties are not, then, potentially infinitely extendable things, as suggested by the second horn of the dilemma.

Consider the idea that the Big Bang is causally related to everything that has and will come after it. Causal ancestors give rise to, but are not thereby identical to, the causal chains that originate from them. One can of course give descriptions of dispositions in terms of sequences of properties, in much the same way as one can give descriptions of the Big Bang in terms of later stages of the causal processes it initiated, but dispositions themselves are no less tractable as discrete entities for such descriptions. This is a familiar point from Donald Davidson's (1980) work on the nature of actions and events. An event may be redescribed in terms of its consequences (my eating an entire mocha cheesecake after lunch may be redescribed as my ruining my appetite for dinner, for example), but this is not to equate the event with its consequences. So the power to bring about the set of causal properties P_1 may well be described as the power to bring about the power to bring about the further set of causal properties P_2, and so on. There is nothing metaphysically problematic in this. And if the concern is shifted to the realm of epistemology – that is, to a concern regarding the ability to attribute causal properties given the regress cited – the problem collapses into one or some combination of the epistemic challenges raised a moment ago, but I have already considered those worries.

My disarming of the second horn of the dilemma may yet leave some feeling uncomfortable. For those who feel it, the discomfort stems, I think, from the fact that DIT entails a kind of holism with respect to the natures of causal properties. This holism is perhaps best described as an ontological circularity. If the identity of a causal property is determined by certain dispositions for relations with other properties, the natures of causal properties taken as a whole are constituted by a vast network of potential relations. The natures of individual properties are thus linked to one another via closed loops of potential relations. Unease about this situation may arise out of doubts about internal relations, the repudiation of which played an integral role in turn-of-the-twentieth-century arguments against idealism during the formative years of analytic philosophy.[3]

---

[3] This was suggested to me by Paul Teller. For an account of this period, see Hylton 1990, e.g. p. 55.

An internal relation is one that is part of the essence of a relatum; the identity or nature of the relatum is thus dependent on the relation obtaining. G. E. Moore thought that the idea of such relations is inherently absurd or contradictory, arguing that in order to assert that a relation obtains at all one must be able to conceive of the relata independently, and certainly not as dependent on the relation itself.

If my arguments in this section have been compelling, however, the ontological circularity of causal properties is not vicious, and furthermore, concerns about internal relations, whatever their merits, are misplaced here. On the dispositional view of properties, no specific relations need obtain in order for causal properties to have their identities. According to DIT, it is simply the potential for relations of various sorts that determines property identity. The identity of a causal property is determined by dispositions that, on the realist account, are genuine properties regardless of whether any particular manifestations come to pass. Thus, property identity does not depend on any particular relations obtaining. It is defined rather in terms of dispositions *for* relations. Dispositions are occurrent properties prior to and independent of any of the particular relations one might use to individuate them in an epistemic context. The plausibility of internal relations is thus a red herring.

5.4 VACUOUS LAWS AND THE ONTOLOGY OF CAUSAL PROPERTIES

Until now, my goal in this chapter has been to elaborate an understanding of causal properties that nicely supports a semirealist account of scientific knowledge. I have sought to clarify a family of concepts including property identity, dispositions, and necessity in establishing the proposed foundations of a plausible scientific realism. We are now in a position to reap some of the fruits of these labours. One of the nicest features of DIT (and an ulterior motive for recommending it) is that it yields an intuitive account of laws. The expression 'law of nature' is a term of art, and different philosophers use it in different ways. In Chapter 6, I will contend that the most common uses of this term are mostly compatible with the metaphysical underpinnings of realism proposed here. In the meantime, let it suffice to say that the shared core of these several uses of 'law' is the idea that law statements describe what entities in the world are like and how they behave, in such a way as to permit generalizations and predictions concerning them. This chapter began by suggesting that one may think of causal laws as relations between causal properties. Some restrict causal-law statements to expressions of the form '*A* causes *B*', but previously I argued that realists should think of causation as a process as opposed to a special relation between events like *A* and *B*. Descriptions of relations between causal properties give causal information, such as information about how altering one property instance will affect others. Let us consider a few aspects of this proposal now.

Earlier I noted that there are different views on the question of what sort of entity causal properties are, in the sense contested by realists about universals, nominalists, and trope theorists. Some believe that one is driven to very specific ontological commitments regarding the nature of properties in order to resolve apparent difficulties associated with the phenomenon of so-called 'vacuous' laws. In this section, I will argue that these difficulties have been overstated. Given an ontological conception of causal laws and DIT, the scientific realist retains a high degree of neutrality with respect to finer-grained ontological disputes about the nature of properties. Of course, one may have other reasons for taking a stand on these deeper metaphysical questions, but the commitments of metaphysics *qua* scientific realism may be satisfied on any one of various rival possibilities. By demonstrating this, I hope to display realism as a more broadly acceptable and widely attractive position than some would suggest.

On an ontological conception, laws of nature are not linguistic entities. One uses natural and mathematical languages to give expression to laws, but laws themselves are part of the fabric of a mind-independent world. Candidates for laws are not things like sentences that can be true or false. Rather, they are possible relations that either do or do not obtain. Relations obtain only if the things they relate exist. Thus, specific causal laws obtain only in worlds containing the requisite causal properties. A vacuous law, as I will use the term, is one that is never actually realized – a relation that never comes to pass. The main worry about vacuity is the idea that if these relations do not obtain, there may be nothing in the world in virtue of which certain law statements and related assertions are true. I will return to the idea of "truthmakers" shortly, but first let us examine the nature of vacuity itself. There are two ways in which it can arise on an ontological conception of laws. Firstly, the relata of the relations that would otherwise make up laws might not exist in the actual world. I will call this the problem of *missing properties*. Secondly, in some cases, though the relata exist, objects with these causal properties might never encounter one another in such a way as to produce a manifestation of the relevant dispositions. Thus, the relations in question are never realized. I will refer to this as the problem of *missing relations*. Let us consider these problems in turn.

The problem of missing properties generates two categories of ostensibly vacuous causal laws: laws that are vacuous in principle, and laws that are vacuous in practice. In-principle vacuity occurs when law statements describe relations between causal properties that do not exist in the actual world for a principled reason. That is not to say that such properties could not have existed if the world had been inhabited by different causal properties, for presumably it is metaphysically possible that such properties exist. Rather, recalling the concept of necessity associated with DIT, it is to say that they are not members of the network of properties found in the actual world. Idealization is a rich source of in-principle vacuity. Consider, for example, the ideal gas law, which as one might expect, ostensibly relates various properties of gases. The corresponding law statement is '$PV = nRT$', and the variables are intended to represent pressure, volume, the number of moles of gas, the universal gas constant, and temperature respectively. Since no gas is ideal, the statement of the ideal gas law is vacuous in the sense that the relevant properties and relations do not exist as described, strictly speaking. (I will argue in the next section that many law statements are merely incomplete as opposed to incorrect, but let us focus on the case of incorrectness here.) The dispositions conferred by causal properties in the actual world are somewhat different from what is described in the statement of the ideal gas law.

The example of the ideal gas law furnishes a nice illustration of both the nature of putative vacuous laws in principle and the way in which they are commonly misunderstood. According to DIT, the identities of causal properties are determined solely by the dispositions they confer, but the dispositions described in this case are toy versions of dispositions in the actual world. Many law statements describe idealizations of this sort. Does this mean that there are no *bona fide* laws of nature to speak of in these cases? An ontological conception of laws diffuses concerns such as these. In common parlance one uses the term 'law' to refer both to relations between causal properties *and* to linguistic devices employed to describe them. But in-principle vacuity attaches, where it does, to expressions of laws, not to laws themselves. The same is true of in-practice vacuity, which I will come to momentarily. Causal laws are never vacuous in principle, given that they are relations between causal properties. Descriptions of laws, however, generally imperfect, often intentionally idealized for practical purposes, are naturally vulnerable to potential vacuity.

Another good example of in-principle vacuity is given by Mellor (1991/ 1980, p. 140) in a discussion of the law governing the vapour pressure of water (the pressure of the vapour evaporated from the surface of a quantity of water). The mathematical formula one uses to describe this law relates vapour pressure to the temperature of water, but is vacuous in principle over some temperature ranges since water is liquid at some temperatures but not at others. One might of course make this expression more accurate by including temperature ranges over which it applies, but one is not always in a position to specify such details. The law itself, or more precisely the laws, since countless relations between determinate (that is, specific values of) causal properties are described by this law statement, are not vacuous. The relevant relations simply do not obtain between determinate properties that are ruled out as a matter of principle given the natures of properties in the world.

In-practice vacuity, the second category of vacuous law statements generated by the problem of missing properties, is also generated by the problem of missing relations. In contrast to the case of missing properties, where vacuous law statements in principle describe causal properties that do not exist for a principled reason, vacuous law statements in practice describe properties that do not exist for reasons that are merely accidental. Imagine that no object has ever had a mass of exactly 1000 kg. Nothing in the natures of the dispositions conferred by masses or any other properties precludes its instantiation, but nevertheless, no object has it. If the expression '$F = ma$' describes a family of laws relating determinate values of force, mass, and acceleration, it is thus vacuous for the specific value $m = 1000$ kg. None of the relations that one might wish to describe in terms of expressions involving this determinate property has ever obtained. The idea that the same kind of situation can be produced by the problem of missing relations is nicely illustrated by a hypothetical scenario conceived by Tooley (1977, pp. 668–9; 1987, pp. 47–8). Imagine that one has grounds for believing that under appropriate circumstances, causal laws would govern the interactions of two fundamental particles that have never and will never encounter each other. Here the requisite causal properties exist, but some of the dispositions they confer are never manifested because appropriate circumstances never obtain. For a summary of how various problems are connected to vacuous law statements, see Figure 5.1.

Understanding the ways in which vacuity in practice can arise, we are now in a position to consider why some think it problematic. The chief concern about in-practice vacuity is connected to the notion that one requires truthmakers for law statements that *would* correctly describe, or for counterfactual conditionals that *do* correctly describe, absent yet possible circumstances. In cases where one has reason to believe that certain never before manifested behaviours are possible, these behaviours are

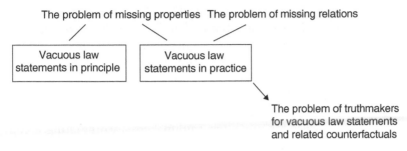

Figure 5.1. Problems associated with vacuous law statements

in principle describable by statements of laws. In the absence of the relevant relations, however, what makes these statements, or assertions of counterfactuals regarding the correctness of these statements, true? As Figure 5.1 indicates, this worry is elicited only by in-practice vacuity. Since in-principle vacuity concerns law statements that are idealized and thus false, strictly speaking, there is no immediate requirement here for truthmakers (though the issue of how idealizations yield knowledge remains; I will consider this important question in Chapter 8). The truthmaker problem is especially acute in the case of in-practice vacuity generated by missing properties, where if one has already formulated a general law expression for a family of relations like '$F = ma$', it is a simple matter to formulate law statements that would apply to specific counterfactual instances, as in the example of $m = 1000$ kg a moment ago.

When in-practice vacuity looms, perhaps the most obvious strategy for supplying truthmakers is to appeal to the existence of transcendent universals. That is, one might assert that the property of mass 1000 kg and all of the relations of which it is capable exist quite independently of concrete objects that may or may not instantiate them. This is Tooley's strategy. All of these relations exist, he claims, in Plato's heaven. A different approach is championed by Armstrong, who believes that all universals are instantiated, or immanent. In cases of vacuous laws in practice, Armstrong claims that counterfactual statements about laws that would obtain if certain properties were instantiated are true in virtue of higher-order laws (for example, the one described as '$F = ma$'), themselves somewhere instantiated. Higher-order law statements give the general form of lower-order laws statements by relating properties of the types cited in the relevant counterfactuals. It is unclear, however, that there will always be the sorts of higher-order laws Armstrong requires to support his counterfactuals. In order that a higher-order law relate the relevant properties *and* be somewhere instantiated, as John Carroll (1987, pp. 271–2) points out, some lower-order law must both relate the same kinds of properties and have the same form as the counterfactually specified, would-be law, and this cannot be guaranteed. Tooley's scenario of the two non-interacting particles, for instance, may furnish a hypothetical example in which no appropriate higher-order law exists.

In any case, whether or not they are tenable, variations of Tooley's and Armstrong's strategies for providing truthmakers, incorporating a commitment to transcendent and immanent universals respectively, are open to the advocate of DIT. Another approach, however, is uniquely available to those holding a dispositional view of property identity. This alternative supplies truthmakers for counterfactuals about law statements that would describe absent yet possible circumstances, if not for the law statements themselves, and in such a way as to leave the matter of whether properties are transcendent universals, immanent universals, or tropes an open question. The aim here is to be able to affirm the truth of certain counterfactuals pertaining to laws that do not obtain because the relevant relations are not instantiated, either because one or more of the properties involved are themselves nowhere instantiated, or because objects instantiating these properties never encounter one another in an appropriate way. According to DIT, there is nothing more to the identity of a causal property than the dispositions it confers. Property identity is exhausted by dispositions for relations with other properties. The *mere existence* of a given property – transcendent, immanent, whatever – thus serves as a truthmaker for any counterfactuals whose consequents correctly describe its possible relations. Thus, if any one party to a possible relation exists, a truthmaker for a counterfactual claim about circumstances under which that relation would obtain also exists!

In fact, the dispositional solution to the apparent difficulty posed by vacuous law statements in practice goes even further than this. The role of truthmaker can be served, not only by any one of the properties potentially involved in the relevant relation, but by *any* property *whatsoever*. Recall that it is a consequence of DIT that networks of causal properties have a holistic nature. This furnishes a more radical solution to the problem of truthmaking than is generally appreciated. The existence of any *one* causal property is a sufficient truthmaker for counterfactuals about all possible relations applicable to the world in which that property is found. This is because, again, the identity of a causal property is determined by its dispositions for relations with other properties. A complete specification of these potential relations thus constitutes an exhaustive set of the possible links between all causal properties. Given that these links map the various ways in which properties by their very natures can be related, this network of properties and potential relations comes as a package or not at all. The existence of any one property, then, is sufficient to determine (ontologically) the natures of all possible properties and relations in the network to which that property belongs.

To invoke a metaphor I have used before, all of the relations of which causal properties are capable are "encoded" in the dispositions they confer. Properties and laws are thus flipsides of the same coin. Appreciating this intimate connection may lead a realist to say one last surprising thing about truthmaking (just in case there have not been enough surprises already). I have argued that on the dispositional account, the mere existence of a causal property serves as a truthmaker for counterfactuals about vacuous law statements, but not necessarily for vacuous law statements themselves. The former are conditionals to the effect that *if* certain properties were instantiated in appropriate circumstances, their instances would stand in certain relations. The latter are statements about relations *simpliciter*, so it is perhaps only natural to suppose that their truthmakers should be the relevant relations themselves. Given DIT and the intimate connection between causal properties and laws, however, one might be tempted to say that uninstantiated laws have a kind of reality, even without recourse to transcendent universals. If one accepts that dispositions are occurrent properties whether or not they are manifested, it may be tempting to say that laws are in a sense actual, whether or not they are actualized. This would amount to the idea, offered here merely as food for thought, that uninstantiated but possible relations "exist" in potential form, standing by to be realized under appropriate circumstances.

5.5 CAUSAL LAWS, *CETERIS PARIBUS*

My consideration of the nature of causal property identity and the account of laws that emerges from DIT is now complete. Before moving on to a discussion of the categories of things realists may take these properties and laws to concern, however, one final piece of business remains to be tidied up. Earlier in connection with the notion of in-principle vacuity, I cited idealization as a common source of expressions that describe relations between causal properties that do not exist as described. Properties conferring such dispositions are not possible members of the network of properties found in the actual world. Though an accurate account of idealization, however, this is too severe a diagnosis of many scientific descriptions. It is certainly the case that one often produces idealized descriptions of the natures of properties, intentionally or otherwise, in order to render them sufficiently tractable (for example, mathematically). However, it is also the case that one often uses descriptions of relations between causal properties that are strictly correct, but only when applied to *different* circumstances. This sort of practice is best described not in terms of idealization, but rather in terms of what I will call abstraction (or more carefully, "pure" abstraction, to exclude cases that are both abstracted and idealized). Idealization and abstraction will both come in for more careful scrutiny in Chapters 7 and 8, but having introduced the former, the latter notion merits brief attention here as well.

Many law statements come with the implicit disclaimer '*ceteris paribus*', 'all things being equal', for one does not expect them to accurately describe the parts of the world to which they are applied in all circumstances. In many situations one uses law statements to make predictions whose accuracy will vary from one circumstance to another. This is because other factors of which no mention is made in these expressions of laws may interfere to produce outcomes that differ from the ones predicted. Cartwright (1989) observes that this predicament is commonly experienced outside of the strict experimental confines in which law statements are generally formulated, such as laboratories. In order to work out the details of a specific relation targeted for investigation, it is often necessary to shield experimental set-ups from interfering factors so as to study the desired relation in isolation. Outside of the experimental arena, however, interfering factors are often plentiful, so one says that the law statements established in the lab are true at best *ceteris paribus*. They describe relations of properties that have been abstracted away from more complex circumstances.

For the dispositional realist, *ceteris paribus* expressions are naturally interpreted as claims about dispositions. These statements describe what happens so long as factors not taken into account do not interfere, and this idea folds neatly into the understanding of properties suggested by DIT.[4] A *ceteris paribus* law statement, if correct, describes relations that obtain in circumstances where no causally relevant properties other than those described are present. If other causally relevant properties *are* present, the dispositions manifested will be different from those whose manifestations

---

[4] See Lipton 1999 for an excellent discussion of dispositions and the semantics of *ceteris paribus* statements, and Cartwright 1989, pp. 190–1, and Woodward 1992, p. 205, on the topic of stable capacities, to which the following discussion is addressed.

are accurately described by the *ceteris paribus* law statement in question. Recall that causal properties are generally many-faceted. They confer not one disposition, but collections of dispositions. Which dispositions are manifested in a given situation will depend on the assortment of causal properties involved. Thus, whether a *ceteris paribus* expression is useful for prediction in different circumstances will in each case depend on how closely it approximates the relations that obtain there. One often uses an abstract law statement as something *like* an idealization – one applies abstractions to situations that are, in fact, more complex than those they properly describe. One does this because abstract law statements are simpler than more detailed ones, and though they may not accurately describe the dispositions manifesting in a given circumstance, they are often "good enough" for predictive and explanatory purposes.

DIT provides an account of what one achieves by accurately formulating *ceteris paribus* law statements to describe causal laws. These expressions are partial maps of property relations. They hold only partially, or *ceteris paribus*, because in formulating them one does not specify all of the potentially relevant dispositions that make up the sets properly associated with the causal properties described. If correct, *ceteris paribus* law statements are accurate descriptions of possible relations between specific causal properties. The presence and absence of other objects and causal properties will determine whether the outcomes predicted by these law statements for specific instances of causal processes are manifested, but the issue of whether predictions match outcomes is irrelevant to the question of whether *ceteris paribus* expressions correctly map *some* possible relations. The relations they describe may be laws of nature regardless.

The idea that causal properties are many-faceted should not be confused with the idea that properties confer different dispositions in different circumstances. Some appear to adopt the latter view (see n. 4), suggesting that the dispositions with which properties are associated may vary from one circumstance to another. If DIT is correct, however, the dispositions associated with specific causal properties are invariant. In order to appreciate the difference between these views it is helpful to distinguish once again between epistemic and metaphysical senses of "association". In epistemic contexts, one does of course associate very specific and often different dispositions with one and the same property in different circumstances. One says, for example, that the molecular structure of a compound (assuming this is a causal property) confers a disposition to dissolve in some situations and a disposition not to dissolve in others. But metaphysically speaking, causal properties are uniquely identified in all circumstances with the same dispositions. To think otherwise is to confuse dispositions with their manifestations. Whether a specific disposition is manifested depends on the presence and absence of other property instances. Possible manifestations such as dissolving and not dissolving are thus tied to circumstances. Possible dispositions, however, depend only on the natures of causal properties.

DIT gives a compelling account of causal properties, which are central to the formulation of semirealism. This thesis also yields a simple understanding of other commonly invoked features of realist discourse, such as *de re* necessity and laws of nature. Earlier I made the surprising suggestion that spelling out these features might not violate empiricist sensibilities much further than my initial portrayal of semirealism had already. Though the idea of *de re* necessity and ontological conceptions of laws are anathemas to empiricism, the account of these metaphysical notions offered here is relatively neutral. Causal laws are nothing more than relations between causal properties. That is, they are nothing more than concrete structures, whether of objects, events, or processes. The ultimate ontological status of the properties and relations making up these structures, whether as universals, sets of particulars, or sets of tropes, is an open question. And the idea of necessity, it turns out, follows simply given DIT, requiring no further ontological ingredients. The weight of metaphysics entailed here, I believe, is substantially less than one might have otherwise supposed.

CHAPTER 6

Sociability: natural and scientific kinds

6.1 LAW STATEMENTS AND THE ROLE OF KINDS

Scientific theories describe causal properties, concrete structures, and particulars such as objects, events, and processes. Semirealism maintains that under certain conditions it is reasonable for realists to believe that the best of these descriptions tell us not merely about things that can be experienced with the unaided senses, but also about some of the unobservable things underlying them. In charting the proposed conceptual foundations of this stance I have covered a lot of ground, from causal processes through *de re* necessity and now to one last crucial concept. There is one item on the inventory of realist commitment about which I have not yet had much to say. In addition to theorizing about and experimenting on instances of properties and relations, the sciences also describe kinds of particulars. Earlier I said that instances of causal properties regularly cohere to form units that are especially apt for scientific study, and it is precisely these groupings that make up the particulars described by theories. The time has come, finally, to consider the particulars of scientific discourse.

The idea of kinds of particulars is perhaps more in need of clarification than any other aspect of the metaphysics of semirealism. For here more than in any other place, scientific realists have allowed their position to become unfortunately entangled with the metaphysical speculations of past, systematic philosophers. Several of these speculations are simply outmoded in the context of modern science and some must now seem defunct even to metaphysicians working outside of this context altogether. This chapter is devoted to thinking about what sort of work the concept of kinds actually does for the realist. By considering its value to a semirealist account of scientific knowledge, a great deal of light is shed, I believe, on the question of how the realist should best think of them. In the long-standing tradition of reflections on this topic, I will speak here of kinds of objects (including countable things, like silkworms, and merely quantifiable things, like silk), though it should be understood that the morals I will draw in connection with objects also apply *mutatis mutandis* to events and processes. With respect to objects, certainly, the more specific invocation of "natural" kinds is common among scientific realists, but accounts of realism rarely give any careful consideration to the important foundational questions: what *are* natural kinds, and what are they good for?

The primary motivation for thinking that there are such things as natural kinds is the idea that carving nature according to its own divisions yields groups of objects that are capable of supporting successful inductive generalizations and predictions. So the story goes, one's recognition of natural categories facilitates these practices, and thus furnishes an excellent explanation for their success. Natural kinds, it is said, have roles to play in scientific theories; a knowledge of natural, mind-independent categories helps to warrant one's inductive practices. For example, one good reason, it might be thought, for believing that an object will turn out to be a certain way (to behave in the manner of some others one has observed, for instance) is that it is bound to be that way, as a member of a kind that *is* that way. Scientific practice can depend on natural kinds. Thus Hilary Kornblith (1993, p. 7) states: 'The causal structure of the world as exhibited in natural kinds ... provides the natural ground of inductive inference.' And Boyd (1999, p. 146) adds: 'It is a truism that the philosophical theory of *natural* kinds is about how classificatory schemes come to contribute to the epistemic reliability of inductive and explanatory practices.' Hacking (2007) traces links between kinds and induction in historical antecedents such as John Stuart Mill, John Venn, C. D. Broad, and Quine.

Given this background, it is not surprising, perhaps, that scientific attempts to codify nature are widely thought of as means to the end of illuminating certain general principles or laws of nature. Kinds of things are, and behave in, certain kinds of ways. To know these ways is to know what the world is like and how it works, and this is to know its laws. What better warrant could there be for inductive practices like generalization and prediction? As I mentioned earlier, however, the expression 'law of nature' is a term of art, and in the present context it will prove helpful to have some understanding of its different uses. I have employed the term 'law' very specifically to denote relations between causal properties, or what I first introduced as concrete structures. The examples of law statements I have used, such as Newton's second law of motion ($F = ma$) and the ideal gas law ($PV = nRT$), can be interpreted straightforwardly as putatively describing relations of this sort. Many supposed examples of law statements concerning natural kinds, however, do not appear to describe laws in this sense. Consider the candidate law statement, 'All planets in solar systems move in approximately elliptical orbits.' This generalization makes no explicit reference to the relations of causal properties. With present purposes in mind, the variety of generalizations commonly referred to as law statements in philosophical discussions can be sorted into three classes. Let us consider them now, in order to clarify their relevance to the notion of kinds.

The first of these classes contains statements I have already examined in some detail, viz. ones that describe relations between causal properties. I will continue to refer to these generalizations as causal-law statements. Generalizations belonging to the second class describe how members of categories of objects behave, as in the example just considered of the statement regarding how planets orbit their suns. Let us call these *behavioural* generalizations. Finally, generalizations belonging to the third class describe the natures of members of categories of objects in terms of one or more of their distinctive or characteristic intrinsic properties. This sort of statement is exemplified, for example, by the claim that 'water is H_2O', in which samples belonging to the category water are described in terms of their composition. Let us call these *definitional* generalizations, since they make partial and sometimes exhaustive reference to properties, the possession of which defines membership in categories of objects. (I will later suggest that functional or other *relational* properties may demarcate kinds as well; in such cases, the relevant generalizations may qualify as both definitional and behavioural.) These three classes of putative law statements each contain inductive generalizations regarding aspects of the natural world, and it is for this reason that they are commonly referred to as law statements. These generalizations are basic tools for scientific practice and everyday life, providing bases for predictions regarding the behaviours and natures of the things to which they apply.

It is unclear, however, whether the ontological conception of laws I adopted earlier is compatible with the third class of supposed law statements, the definitional generalizations. In the case of causal law statements and behavioural generalizations, there are relations and regularities in the world in virtue of which these statements are true (or close by, so the realist hopes). Here the idea that laws are parts of nature as opposed to linguistic entities is viable. Definitional generalizations, however, simply describe categories of objects in terms of their properties. The relevant truthmakers for statements of this sort are objects of the relevant kinds, and it rather strains the senses of the terms 'object' and 'law' to claim that objects are laws! On an ontological conception of laws of nature, definitional generalizations are not law statements. They are merely descriptions of objects. 'Water is H_2O', 'mammals are warm-blooded', 'electrons have negative charge', etc., are simply statements that describe categories of objects in terms of their properties. They are, nevertheless, useful for purposes of prediction and explanation, and this is usually all that is intended by those who refer to them as law statements. Given that they are dubious candidates on an ontological conception of laws, I will speak of them (together with behavioural generalizations; see Table 6.1) simply as law-like generalizations instead. Each of these cases is subject to the hypothesis that natural kinds facilitate inductive generalizations about what entities in the world are like and how they behave, on the basis of which scientific practices like prediction succeed.

The entities described by causal law statements are causal properties and their relations. In developing an account of property identity, Chapter 5 laid the groundwork for a concept of property kinds. Kinds of causal properties are categories whose members confer the same sorts of dispositions. For example, families of properties such as masses, temperatures, optical densities, and so on are kinds of properties in this sense. In this chapter I am concerned with kinds of objects; consequently, behavioural and definitional generalizations will take centre stage initially. Causal laws, however, are never far away. Wherever there are regular behaviours such as orbitings of suns by planets there are causal processes, and wherever there are causal processes there are causal laws. As dispositions conferred by the properties of objects are manifested in appropriate circumstances, these laws entail regularities. Behavioural generalizations usually make no mention of the underlying relations of properties that produce the regularities they describe, but they are there nonetheless. And though the features of the world described by definitional generalizations are (presumably) not always entailed by causal laws, they are never wholly unconnected. For definitional generalizations are descriptions of objects, and objects are cohering sets of instances of causal properties.

The goal of this chapter is to examine the connections suggested here between kinds, law-like generalizations, and epistemic practice, and to consider how the scientific realist should best understand them. Given that various facts about the natures and behaviours of kinds of objects are described by definitional and behavioural generalizations, respectively, and the currency of the common realist assumption that natural kinds underwrite some of the epistemic functions generally thought to involve

Table 6.1. *Three types of law-like generalizations*

Types	Content
Causal law statements	Descriptions of relations between causal properties
Behavioural generalizations	Descriptions of behaviours of kinds of particulars
Definitional generalizations	Descriptions of the natures (distinctive or characteristic properties) of kinds of particulars

such generalizations, there is a *prima facie* case for examining these connections. What is it about natural kinds, exactly, that is supposed to qualify them for the job of "underwriting"? The plausibility of this picture ultimately depends on the details of the account the realist gives of their relation to generalizations and predictions. A moment ago I accused those who offer versions of scientific realism of paying insufficient attention to these matters, but there is a notable exception to this charge. In recent years, Brian Ellis (1999, 2000, 2001) has attempted to address precisely these issues, by developing an account that he and others call the New Essentialism (NE). This position connects a view of kinds, conceived as having essences, with law-like generalizations, in such a way as to furnish the details I suggest require scrutiny if one is to allow natural kinds a place in the ontology of realism. NE analyses the relation of kinds to inductive success by linking a dispositional view of the essences of kinds to the generalizations that figure so prominently in scientific practice.

In the rest of this chapter, using NE as a springboard, I will endeavour to furnish a compelling account of kinds for the semirealist. Along the way I will argue that NE, and indeed any view that admits as genuine only kinds with essences, does not adequately explain the link between kinds and inductive success. Scientific practices such as generalization and prediction are concerned not merely with kinds having essences but also with kinds lacking them, and this presents a difficulty. If one's account of the connection between kinds and epistemic practice is premised on the idea that kinds have essences, one is left without a connection in the case of kinds lacking them. As we shall see, practices such as generalization and prediction are facilitated by causal laws and distributions of causal properties. Kinds of objects are the subjects of helpful generalizations derivatively, in virtue of having these properties. Behavioural and definitional generalizations describe patterns that obtain in the world, to the extent that they do, because of the ways in which causal properties are distributed among conventionally demarcated categories of objects. The proposed semirealist understanding of the kind concept, I will suggest, makes sense of *all* scientific kinds. By throwing off the yoke of an antiquated metaphysics, this proposal leads to a deflationism and a pluralism regarding kinds of objects that is appropriate to the sciences today.

6.2 ESSENCES AND CLUSTERS: TWO KINDS OF KINDS

Like 'law of nature', 'natural kind' and 'essentialism' are terms used in many ways by different people, and not all of the various connotations with which they are associated will concern me here. Two of them are fundamental, however, and cannot be left aside. Natural kinds are said to be objective. They are nature's own divisions, not ones that are merely useful, convenient, or of interest to humans. Furthermore, it is traditionally held that what makes a member of a natural kind a member is the possession of a kind essence – a set of intrinsic properties. To be lacking any of these properties is to preclude membership and *vice versa*. The possession of these properties is necessary and they are jointly sufficient for kind membership. Combining the ideas of objectivity and essence, one can formulate Locke's (1975/1689, Book III, ch. III, §15) distinction between real and nominal essences, the former constituting objective categories and the latter merely the categories one uses. Realists with respect to both scientific knowledge and kinds generally hold that our best theories provide descriptions of real essences, and this is certainly the view of NE. The connotations of objectivity and essence will be my focus here. Other common connotations, such as the supposed immutability or atemporality of kinds, usually associated with Aristotle, and the idea that essences are generally composed of underlying microstructural properties, reintroduced into recent discussions by Kripke and Putnam, are not central to a plausible conception of kinds, or so I will suggest.

NE is a convenient place to start, for one of its chief concerns is to offer an account of the relationship between kinds and law-like generalizations that incorporates the notions of objectivity and essence. Ellis's primary motivation is to oppose a Humean world view, according to which, he says, laws are 'imposed' on otherwise passive, causally indifferent objects. Consider, for example, the inert corpuscles of seventeenth-century natural philosophy, whose behaviours are described and somehow determined from without by Boyle's and Newton's laws. NE, conversely, views laws as immanent in the causal powers of objects: 'Laws of nature depend on the essential properties of the things on which they are said to operate, and are therefore not independent of them' (Ellis 2001, p. 1). On this view, law-like generalizations describe the essential properties of the members of kinds, and these properties are conceived as powers or dispositions. As the assiduous reader will have guessed, I am sympathetic to Ellis's motivation. Many realists have reasons for not wanting to be thoroughly Humean, and I have discussed some of these reasons earlier in the contexts of causation, dispositions, and necessity. But Ellis formulates his opposition to Hume in an unfortunate way. For one thing, NE adopts too narrow a conception of kindhood, acknowledging as genuine only kinds with essences. As we shall see, law-like generalizations only sometimes describe the essential properties of members of kinds, and frequently do not.

The most obvious and compelling sources of resistance to an exclusive commitment to kinds with essences are the sciences themselves. The kinds of objects investigated by the sciences are sometimes describable in terms of essences, but often resist this sort of description. The traditional view that kinds are ontologically distinguished by essences has a storied past, but many of the kinds one theorizes about and experiments on today simply do not have any such things. Many of these kinds are groups whose members need have *no* distinguishing properties in common, and this clearly violates the stipulation that essences comprise sets of properties that are necessary and jointly sufficient for kindhood. I will refer to kinds with essences and those without as *essence* kinds and *cluster* kinds, respectively. Canonical examples of essence kinds are familiar from physics and chemistry. The kind essence of an electron, for example, consists in a handful of determinate, state-independent causal properties (specific values of mass, charge, and spin) that are characteristic of all and only members of this kind. But not all kinds fit this model.

The best-known examples of cluster kinds are derived from attempts to explicate the species concept in biological taxonomy. It is generally agreed that the search for essences here has failed. For example, neither morphological nor genetic properties will do, due to intra-species variation and overlap with other species. Reproductive isolation is also often cited as the mark of a species. Imagine that such isolation could be accounted for in terms of sets of intrinsic properties shared by certain individuals which unite them reproductively and isolate them from others. This proposal is also inadequate to the task of specifying essences, for several reasons: hybridization violates reproductive isolation, and when it occurs offspring are sometimes fertile, thus compounding the problem; some subpopulations within species mate successfully with other sub-populations but not with all; focusing on these sorts of reproductive criteria ignores asexual species entirely. Furthermore, in keeping with both intuition and biological practice, membership in a species cannot be conceived in terms of necessarily possessing distinctive morphological or reproductive properties (that are jointly sufficient), for a sterile tiger would still be a tiger, as would a tiger with only three legs, or an albino. I will consider the different concepts of species in the next section, but for now let it suffice to say that none of them identifies species with essences as traditionally understood, in terms of intrinsic properties that are both necessary and jointly sufficient for membership.

Given the absence of kind essences for various things widely regarded as kinds, it is now common to relax the essence criterion in the demarcation of many scientifically sanctioned categories of objects. In such cases membership in a kind is usually described in terms of metaphors: clusters, family resemblance, or as Hacking (1991, p. 115) puts it, 'strands in a rope'. These are polythetic kinds, meaning that the possession of a clustered subset of some set of properties, no one of which is necessary but which together are sufficiently many, entails kind membership. NE, which endorses the traditional appeal to essences in distinguishing kinds, is not surprisingly uncomfortable with cluster kinds. If the essences of kinds underwrite the success of scientific generalizations and predictions, then it is unclear how scientists and others meet with success in these practices in the case of cluster kinds. And if this is so, the kind concept has little or dubious utility for the scientific realist in these cases. How is one to square this with the fact that the sciences routinely theorize about and experiment on kinds without essences?

It would appear that realists face a dilemma here. On one horn they can insist that only essence kinds are genuine kinds, but then face the task of explaining away the kinds investigated by scientific disciplines whose subjects are clusters, and nonetheless admit of successful generalization and prediction. NE embraces this first horn of the dilemma. Inductive generalizations describe kinds with essences, and it is the existence of essences that fosters scientific success. But what then of the cluster kinds of parts of biology and perhaps other scientific disciplines? On the other horn of the dilemma realists can accept that there are both essence and cluster kinds, but then face a challenge to account for the connection between cluster kinds and successful epistemic practice. This challenge is easily met in the case of essence kinds, for as well as being jointly sufficient, the properties that compose essences are each *necessary* for membership. Since every member of an essence kind has certain properties necessarily, generalizations regarding the presence of these properties or the behaviours to which they give rise are guaranteed to be law-like, thus promoting successful inductive practices. In the case of cluster kinds, however, members need not possess *any* specific property in the set loosely identified with the relevant clusters, so there is no guarantee that generalizations concerning them will be law-like at all.

Consider the first horn, to which NE is committed. There are two readily apparent strategies for grasping it. T. E. Wilkerson (1995, p. 132) and Ellis (2001, p. 21) advocate one of them when they claim that biology does in fact deal with essence kinds after all. Though the populations generally regarded as species by biologists are not kinds *per se*, they *are* groups of closely related kinds, whose essences are genetic constitutions. For example, human beings (*Homo sapiens*, say) are not all members of the same kind, because they have different genetic make-ups. What one calls human beings are objects with closely related genomes (sets of genetic material). Let me generalize this strategy on behalf of NE: to explain away a cluster kind in favour of essence kinds, refine the search for essences until acceptable candidates emerge. In the case of biological species this process of refinement is concluded, ultimately, in the genomes of individuals. But this strategy defeats itself, for the process of refinement in search of essences has a cost. In the course of refinement, the extensions of the relevant kinds are dramatically reduced. With respect to biological kinds, for example, such essences would almost always be instantiated by unique individuals, with relatively rare exceptions in cases such as identical twins and clones. So much for the connection between kindhood and successful generalizations and predictions! It would be a strange biology indeed whose epistemic practices were confined primarily to specific individuals.

A second possible strategy for those inclined to grasp the first horn of the dilemma is to explain cluster kinds away by invoking a form of reductionism. One might think of essence kinds as the building blocks of the natural world. The natures and behaviours of other, cluster-type, so-called "kinds" could then be analysed and understood in terms of different combinations of their more basic constituents. In this way, one might hope to retain an exclusive commitment to essence kinds and meet the challenge of explaining how and why the sciences acknowledge, theorize about, and experiment on other "kinds" of things. Now, perhaps there are such things as fundamental or otherwise basic essence kinds. Perhaps they comprise things such as the particles of current subatomic physics or more elementary entities yet unknown. Whatever the case may be here, the idea that the success of inductive generalizations and predictions regarding cluster kinds can be explained in terms of the natures and behaviours of essence kinds is ultimately untenable. This strategy for grasping the first horn of the dilemma fails, not only because it is uncertain whether cluster kinds are reducible to essence kinds, but also because it is doubtful it would help if they were. Let us consider why this is so.

There are several ways of thinking about reductionism, but for present purposes all that is required is a broad distinction between ontological and explanatory reduction. At first glance ontological reduction may seem trivial. After all, what controversy could there be in the claim that larger things are composed of arrangements of their components? Precisely this claim, however, is doubted by those who maintain that wholes are in some cases greater than the sums of their parts, and that organized systems have emergent properties that are not mere combinations of the properties of their constituents.[1] The debate concerning ontological reduction has immediate implications for the prospects of explanatory reduction. Those who subscribe to the triviality of ontological reduction may claim that the natures and behaviours of clustered wholes are in principle explainable in terms of their essence-kind parts. On this view it is possible, in principle if not in practice, to explain the behaviours of acids and elephants in terms of subatomic particles or perhaps some other less basic essence kinds. If there are such things as emergent properties, however, then the natures and behaviours of larger things generally are not explainable in terms of more basic essence kinds, even in principle.

In practice, one is concerned not with the mere possibility of explanatory reductions but with formulating them, and this sort of achievement is not a widespread feature of scientific work. Zoologists do not make generalizations and predictions about elephant behaviour in terms of subatomic particles, and not because they are lazy (the zoologists, not the elephants). The only definitive evidence for ontological reduction is comprehensive explanatory reduction, but since the latter features little in scientific practice, it remains an open question whether the natures and behaviours of cluster kinds are reducible to those of essence kinds. And even if it were the case that ontological reduction applies universally to cluster kinds, it is unclear how this would help NE to meet the challenge it faces on the first horn of the dilemma. For even if token elephants can be exhaustively decomposed into essence kinds, that would not by itself explain why

---

[1] For discussions of emergence, see Beckermann, Flohr, and Kim (eds.) 1992 and Schrödinger 1967, ch. 7. See also Mellor and Crane 1991/1990, p. 87, for examples from physics that defy the notion of explanatory reduction to component parts. As the authors point out, physics is sometimes *macro*reductive.

generalizations and predictions about the *kind* 'elephant' are successful, to the extent that they are. That is to say, even assuming ontological reductionism, it would remain to be explained why populations of organisms lacking essences support successful inductive practices, because if a kind has no essence, this implies that its members are composed of different essence kinds. Thus, even if token clusters-that-are-elephants are ontologically reducible to essence-kind constituents, this would not entail that inductive success regarding the *category* of elephants is amenable to explanatory reduction.

It is worth making a final, perhaps less compelling but nonetheless suggestive point about requiring essences as a criterion of natural kind-hood. One might argue that this standard results in too revisionist an ontology of kinds. Essence kinds make up some but not all of what are recognized as kinds of objects across the sciences. If one accepts NE or any other view holding that kinds must have essences, one must surrender the others. Of course, there is a sense in which this consequence hardly matters. The rest of physics, and chemistry, biology, and the applied sciences, neither crave nor require the metaphysical validation of being concerned with genuine natural kinds. The issue runs deeper than this, however. Different scientific disciplines investigate different categories of natural phenomena, and in giving an account of scientific knowledge, it seems only reasonable that realists should view these different subject matters as constituting naturally specifiable divisions. If this is so, then essences are no exclusive desideratum in the demarcation of kinds. There is something wrong with a theory of kinds that is relevant only to certain scientific specialties, and more importantly, since other specialties are perfectly good at furnishing successful generalizations and predictions, it is a mistake to link the utility of the kind concept to essence kinds alone.

A moment ago I suggested that realists face a dilemma. Either they accept only essence kinds, in which case explaining scientific success in terms of natural kinds is simple, but only in some areas, and at great cost to one's stock of kinds; or they accept both essence and cluster kinds, but then must provide an account of how knowledge of the latter facilitates successful inductive practices. NE attempts to embrace the first horn of this dilemma, and as a consequence, I have argued, impales itself. I believe the semirealist should instead embrace the second horn and seek a place for all manner of kinds in the world view of scientific realism. Before embarking on this task, however, one final objection to the idea that cluster kinds are the subjects of successful generalizations and predictions needs attention. My paradigm case of a cluster kind thus far has been a biological species, but there are several conceptions of species employed in contemporary biology, and some argue that species are not things that admit of inductive generalizations at all. This is exemplary of the more general claim that biological sciences are not law-like or predictive, and if that were so, it would be open to the advocate of NE to maintain that there is no need to connect cluster kinds with inductive success in the first place. Let us consider this possibility now.

6.3 CLUSTERS AND BIOLOGICAL SPECIES CONCEPTS

The idea that biological species are natural kinds is not universally accepted, but those who reject it usually do so because they believe that species are not correctly described by certain traditional connotations of the term 'natural kind'. The most off-putting of these are the requirement that members of a kind share an essence, and the notion that kinds have eternal, immutable natures. Both of these problematic connotations, however, are dispensable. The former is relevant only to essence kinds, and the latter, though indeed part of some past conceptions of kinds, is not a necessary feature of either essence kinds or cluster kinds as I have described them. In this section I will consider in more detail the idea that in many cases, species can be regarded as cluster kinds in the context of contemporary biology.[2] There is no one concept of biological species. In fact, there are several, each of which suits a different combination of scientific goals and explanatory purposes. Traditionally, it is a matter of controversy whether these concepts pick out groups of objects (organisms) suited to law-like generalizations, supporting inductive projections with respect to their members. I will argue that not only are there cluster kinds in biology, but that theorizing and experimentation concerning both species and biological kinds more generally strongly suggest that generalizations and predictions are hardly rare in the biological sciences.

The *phenetic* species concept demarcates groups of organisms by measuring how similar they are to one another, as determined by calculations based on their shared phenotypic traits (the detectable properties of organisms manifested in causal processes involving their genomes, *inter alia*). The sorts of reproductive criteria I mentioned earlier are derived

---

[2] Ruse 1987 defends the idea that species are kinds. Ghiselin 1974 and Hull 1976 and 1978 disagree, arguing that species are spatiotemporally extended individuals. Kitcher 1984 and Dupré 1993 argue for a pluralism of accounts. See also de Sousa 1989 for a critical discussion of the debate. For a detailed survey of species concepts, see Ereshefsky 2001, pp. 80–93.

from the *interbreeding* species concept, according to which species are distinguished by the memberships of individuals in actually or potentially successful interbreeding populations. On the *ecological* species concept, members share an ecological role or niche. Finally, the *phylogenetic* species concept appeals to evolutionary history, demarcating species as lineages whose boundaries are determined historically by instances of speciation and extinction. Two centrally important questions emerge here in the present context. Do any of these concepts of species demarcate cluster kinds, and if so, are the members of these kinds the subjects of generalizations and predictions?

The phenetic approach classifies organisms according to their shared intrinsic properties, and since these classifications are made on the basis of overall similarity measures as opposed to sets of properties that are necessary and jointly sufficient for membership, the idea of cluster kinds seems very appropriate here. It is doubtful, however, whether a convincing case for cluster kinds and generalization and prediction in biology can rest on theorizing in the phenetic tradition, for the approach is not widely accepted. In order to allow practicable calculations of overall similarity measures, some traits must be given more weight than others, and different weightings produce different and incompatible taxonomies. In the absence of biologically significant reasons for adopting any one specific convention regarding weighting, the phenetic approach to the species question has proven less attractive than its cousins.

Let us move on, then, to consider the interbreeding and ecological approaches. A striking feature of both is that their criteria for species demarcation appear to involve relational properties as opposed to intrinsic properties. On the interbreeding concept, for example, it is a relation, or rather some set of relations between individuals concerning their membership in one and the same interbreeding population, that identifies them as belonging to the same species. Similarly, on the ecological concept, relations between organisms and environments distribute the former into separate groups. The suggestion here that kindhood might be determined by relational properties appears to violate the stricture shared by traditional views of kinds and NE, according to which essences consist of intrinsic properties. It may be possible, however, to recast the description of these species concepts in terms of intrinsic properties. As I suggested earlier, for example, one might analyse reproductive isolation in terms of sets of intrinsic properties shared by certain individuals which unite them reproductively and isolate them from others. On both the interbreeding and ecological approaches, a case could be made for a shift in emphasis from relational properties that determine kindhood to intrinsic properties that confer dispositions for these relations.

But even if one were to accept this shift, or alternatively, simply admit relational properties as possible essences, it is doubtful that one would regard interbreeding and ecological kinds as essence kinds. It seems unlikely that even an ardent proponent of one of these approaches would agree that an organism failing to stand in a specified relation, or lacking a disposition that is *ex hypothesi* part of its kind essence, should be excluded from membership in the species to which one would otherwise judge it belongs. If I lacked the intrinsic properties that would allow me to reproduce successfully (in appropriate circumstances), it is unlikely that even Ernst Mayr, who famously championed the interbreeding species concept, would judge me to be something other than a member of *Homo sapiens*. Similarly, those adopting the ecological species concept would likely not exclude an organism that, for whatever reason, performed its "assigned" ecological role poorly or not at all, and lacked the requisite dispositions to do so. This suggests that interbreeding and ecological species classification are not to be conceived on the model of essences, and that cluster kind concepts are lurking here in the background. Thinking of kind membership in terms of clusters of properties allows one to apply common sense in the classification of organisms that fail to have or to manifest the dispositions for breeding or ecological roles normally associated with members of the kind.

Phylogenetic approaches also appeal to relational properties in order to demarcate species. Unlike the case of the interbreeding and ecological approaches, however, phylogenetic criteria cannot be recast in terms of the intrinsic dispositions of the members of species. There is no analogous shift in emphasis possible with respect to phylogenetic properties, such as the relational property of belonging to a specific historical lineage. These relational properties are not unconnected, of course, to the intrinsic properties of organisms having them, but species demarcation on the basis of phylogenetic properties cannot be described in terms of such intrinsic properties. For though there is an explanatory relationship between these intrinsic and relational properties, the direction of explanation here is one-way. The presence of many of the intrinsic properties of organisms can be explained, in part, by appealing to the relations of these organisms to their ancestors, but not *vice versa*. No doubt some facts about an organism's historical relations can be *inferred* from a knowledge of its intrinsic properties (in combination with other data), but an organism's intrinsic properties do not explain its relations to the past! The principle of species demarcation on phylogenetic approaches is thoroughly relational, and this disqualifies phylogenetic groupings as essence kinds according to traditional views and NE.

Nevertheless, Paul Griffiths (1999) argues that the relational property of having the historical origin of a particular lineage constitutes a non-traditional, 'historical essence': the property of being 'a member of the genealogical nexus between the speciation event in which the taxon [the category of organisms] originated and the speciation or extinction event at which it will cease to exist' (p. 219; cf. Sober 1980). Despite flouting the traditional condition that species essences comprise intrinsic properties, historical essences do fulfil the further requirements of essences, namely that they comprise properties that are necessary and jointly sufficient for kind membership. Furthermore, Griffiths suggests that historical essences can help to underwrite successful inductive generalizations and predictions. There are Darwinian grounds, he says, for expecting that taxa whose members share a common descent will also share morphological, physiological, and behavioural properties, as a consequence of what he suggestively calls 'phylogenetic inertia'. This admits of two forms. There is an "Aristotelian" variety, in which forces of natural selection are required in order to maintain the relevant traits in a population, and a "Newtonian" variety, in which traits are maintained with indifference to their adaptive utility. The analogy here is to the inertial motion of bodies in Aristotelian and Newtonian physics. Inertial motion requires the continued application of a force in the former, but not in the latter.

If a significant number of intrinsic properties (morphological, physiological, behavioural) are likely to be retained throughout the duration of a phylogenetically determined lineage, this would indeed help to promote the success of generalizations and predictions regarding the members of such a lineage. Note, however, the bait and switch! We began with the idea that members of a phylogenetic kind share a relational essence, but the properties that would underwrite successful inductive practices here are not the ones making up such an essence. That is, they are not necessary properties of organisms belonging to a phylogenetic species. Rather, they are the intrinsic properties that members of such a kind may possess but need not, depending on whether and to what extent phylogenetic inertia has maintained them throughout the duration of the relevant lineage. To the extent that inductive practices are facilitated by phylogenetic classifications, it is because these kinds also happen to be cluster kinds, not because their members share historical essences *per se*. Sharing a historical essence is consistent with having many intrinsic properties in common, but it is also consistent with considerable diversity. Some phylogenetic kinds show significant alterations in the intrinsic properties of their members over time, and in these cases there may be little scope for practices such as enumerative induction, even though the members of these kinds may be described as sharing historical essences.

There are two important morals to be extracted from the preceding discussion of biological species concepts. The first is that there are in fact cluster kinds in biology, even if it is not the intention of any particular species concept to identify clusters as a means to the end of species demarcation. Secondly, in cases where one is able to use generalizations about species to facilitate inductive projections regarding their members, success is grounded in the fact that there are shared intrinsic properties, not shared essences whether intrinsic or relational. I will consider the kind–success relation in more detail shortly. Before this, however, there is one last point to be made in connection with biological kinds, for as I mentioned earlier, some argue that biology is not a discipline much concerned with generalizations and predictions at all. If such practices are atypical of the biological sciences, then there is little need to explain a supposed connection between biological cluster kinds and these forms of inductive success. This would weaken the case against views such as NE by diminishing the evidence arrayed against the claim that the success of these practices can be accounted for solely in terms of generalizations regarding essence kinds.

Samir Okasha (2002) notes that the interbreeding, ecological, and phylogenetic species concepts are formulated in such a way as to differentiate kinds that play interesting and important roles in evolutionary biology. Though he is pluralistic about these concepts in particular, he goes on to endorse a surprising orthodoxy regarding the nature of biology as compared to other natural sciences: 'Classification in biology, unlike in chemistry, is not concerned with causal generalisation, but rather with identifying those units that play a fundamental role in the evolutionary process' (p. 209). In some ways, perhaps, the prevalence of this orthodoxy is not difficult to understand. It is often cited, for instance, as furnishing an example of the manners in which biological methods differ from those of the physical sciences. Furthermore, the tremendous importance of phylogenetic relationships to some of the most celebrated investigations of modern biology is no doubt responsible, in part, for the lasting presence of this view. Mark Ereshefsky (1992, p. 688) succinctly articulates the importance of phylogeny:

Since the inception of evolutionary theory, species taxa have been considered evolutionary units, that is, groups of organisms capable of evolving. The evolution of such groups requires that the organisms of a species taxon be connected by hereditary relations. Hereditary relations, whether they be genetic or not, require that the generations of a taxon be historically connected, otherwise information will not be transmitted. The upshot is that if species taxa, or any taxa, are to evolve, they must form historically connected entities.

There is no disputing the centrality of evolutionary biology and phylogenetic relationships to modern biology as a whole.

Does this suggest, however, that biology as a discipline is uninterested in kinds *qua* categories whose members admit of inductive generalizations and predictions? As it turns out, the answer is no. Two simple considerations help to demonstrate this. Firstly, there is an important difference in scope, as Okasha observes, between the interbreeding and ecological species concepts on the one hand and phylogenetic concepts on the other. The application of the former generally results in categories whose members are groups of relatively contemporaneous organisms, whereas the latter generally pick out categories whose members exist over significantly longer stretches of evolutionary time. Not surprisingly, more or less contemporaneous members of a given species typically share significant numbers of intrinsic properties, thus promoting successful generalizations and predictions. Indeed, not merely are such practices possible, but many biological subdisciplines *rely* on this fact, for their investigations are premised on the efficacy of these inductive practices, and this leads to the second point in favour of generalizations and predictions in biology.

The centrality of evolutionary theory and the fascinating issues surrounding its conceptual foundations tend to obscure everyday, pedestrian facts about the biological sciences, all too obvious to its practitioners but glossed over by those primarily interested in the nature of evolution. These pedestrian facts are reflected in the truism that all of biology is not evolutionary biology, and not all biological kinds are species and higher taxa. In arguing for pluralism with respect to biological kinds, Kitcher (1984) employs a distinction credited to Mayr between functional biology and evolutionary biology, the former being concerned with 'proximate' causes, and the latter with 'ultimate' causes, of things such as morphology and behaviour. The distinction is useful, I think, in illuminating the fact that much of biology concerns questions at a certain distance from evolutionary processes, and when it comes to these questions, kinds whose members facilitate generalizations and predictions are paramount.

Consider, for example, the explanatory aims of physiology, functional anatomy, and comparative anatomy. In these fields one routinely studies dispositions conferred by causal properties on the members of various kinds of objects, such as organs, organ systems, and organisms, resulting in generalizations and predictions. The same can be said about the categories of things that cell biologists, tissue biologists, and immunologists theorize about and investigate experimentally. In many, specifically goal-directed biological activities, such as medicine, ecosystem management, and population control, inductive projections are crucial. So not only are there cluster kinds in biology, but it is simply incorrect to say that the biological sciences are not much concerned with inductive generalizations and predictions. It would appear that views of natural kinds according to which the success of these practices can be accounted for wholly in terms of essence kinds are still, as I argued earlier, in a bind. Let us return now to the more general question of how to connect kinds with successful epistemic practice.

6.4 SOCIABILITY (OR: HOW TO MAKE KINDS WITH PROPERTIES)

Recall once again the dilemma facing realists, which arises in the context of the attempt to explain the connection between practices of classification and the epistemic success associated with law-like generalizations. On the first horn of the dilemma one holds that only essence kinds are genuine natural kinds, but then must explain away the kinds belonging to scientific disciplines that theorize about and experiment on clusters. On the second horn, one accepts that there are both essence kinds and cluster kinds, but then must give an account of successful inductive practice in the latter case where there is no recourse to essences, and thus no guarantee of law-like generalizations afforded by properties whose presence is necessary. NE embraces the former horn of this dilemma, but any approach that does so, I have argued, gives an inadequate account of the relation between kinds and inductive success. It is now time to grasp the second horn, and to decide whether the natural kind concept can be understood in such a way as to earn a place in the ontological commitments of scientific realists.

So let us start again, this time with the assumption that both essence kinds and cluster kinds are genuine, and that both sorts of classification facilitate successful inductive practices. Is there something that deserves credit, metaphysically speaking, for supporting scientific generalizations and predictions in both cases, not merely one or the other? Consider the essence kinds. What explains their predictive regularity? With respect to certain essence kinds in physics, for example, physicists have identified properties such as mass, charge, and spin, in virtue of which these basic constituents of matter are thought to interact. By now I hope it is well understood what I mean by 'in virtue of' here. These are causal properties. They confer dispositions for behaviour on the objects that have them. It is because subatomic particles have the properties they do that general-izations about their natures and behaviours, where true, hold. But note, this is no less true of acids and elephants. Causal laws relate the *properties* of things regardless of whether they belong to essence kinds or cluster kinds. Both traditional views and NE attempt to ground law-like beha-viours in the essences of kinds, but what explains the fact that these generalizations obtain is not, in the first instance, the fact that some objects have essences. This is where NE goes astray. Law-like behaviours obtain not merely as a consequence of the possession of essential properties by members of essence kinds, but as a consequence of the possession of *any* causal property by *any* sort of object.

The question of whether a property is possessed essentially by a member of a kind is irrelevant to the relations of which it is capable in virtue of having that property. NE, however, has a response to this contention. Wilkerson (1995, pp. 33–4) and Howard Sankey (1997) argue that although one might think it a mistake to ground inductive success in essence kinds, given that one makes excellent inductions about other things as well, this is to misunderstand the metaphysical facts of the matter. One makes reasonable inferences, for instance (taking Sankey's example), about the dangers of fast-moving automobiles, which are clearly not essence kinds. But these sorts of inductions are reliable, he says, *because* they are based on facts about the membership of certain objects in essence kinds – in this case, not automobiles, but a more generic kind of object whose members have relatively large mass and high velocity. I submit, however, that nothing worth having here is gained by the reification of an essence kind tailor-made to save NE. It is not facts about the essence of an imagined kind consisting in heavy, fast objects that underwrites one's inductions here, but rather facts about causal properties such as masses and velocities. Scientific law statements often do not make reference to kinds of objects at all, but rather focus on the causal properties of objects. What sorts of kinds have these properties, whether they are essence kinds or cluster kinds, natural or artificial, is a separate matter entirely.

If this is the case, however, why retain the concept of natural kinds at all? Perhaps the very idea is a vestige of an outdated metaphysics. Perhaps not only empiricist sceptics but realists too should dismiss them as belonging to a bygone era. This, I believe, would be giving away too much. There is an important feature of what realists take to be a mind-independent reality that is not captured by the notion that the dispositions conferred by causal properties account for the regular behaviours of things. Properties, or property instances, are not the sorts of things that come randomly distributed across space-time. They are systematically "sociable" in various ways. They "like" each other's company. The highest degree of sociability is evidenced by essence kinds, where specific sets of properties are always found together. In other cases, lesser degrees of sociability are evidenced by the somewhat looser associations that make up cluster kinds. In either case, it is the fact that members of kinds share properties, to whatever degree, that underwrites the inductive generalizations and predictions to which these categories lend themselves. This is a reflection of the striking, poetic fact that some collections of property instances like each other's company and others do not. It is this fact that one captures with talk about natural kinds, and this feature of reality surely has a place in the ontology of scientific realism.

Most of what I earlier called law-like generalizations regarding the members of kinds are in fact parasitic on laws relating properties. These are not properties of kindhood, such as that of 'being an electron' or 'being an elephant', but quantitative and determinate causal properties, on the basis of which the sciences construct taxonomies to begin with. Given that both essence kinds and cluster kinds have, to different extents, sociable intrinsic properties, have I answered the challenge of the second horn of the dilemma for realists, viz. that of connecting natural kinds with inductive success? The answer, it seems, is yes. In the case of inductions concerning behaviour, dispositions conferred by properties for various manifestations are present wherever such properties are found, and to the extent that the same causal properties are found in members of the same kind, their behaviours will be subject to inductive generalizations and predictions. Inductions concerning the natures of kinds, such as the characteristic compositions of their members, will likewise reflect the distributions of properties within them. Any member of a category of objects that shares causal properties with other members, either strictly in the case of essence kinds, or more loosely in the case of cluster kinds, can be expected to be similar and to behave in similar ways in similar circumstances. One's expectations in this regard are appropriately shaped by how strict or loose a kind one is considering.

Sociability, of course, is just a metaphor, intended to describe the metaphysical fact that in cases referred to as examples of kinds, property instances tend to cluster. In the upper limit of sociability the properties composing the sets definitive of kinds are necessary and jointly sufficient for membership. That is, they constitute what one calls "essences". No doubt the idea of sociability can be analysed further, but it is doubtful whether any one analysis will apply to kinds across the board. Boyd (1999), for example, analyses it in terms of what he calls 'homeostatic clustering'. Homeostasis is understood here as a product of causal mechanisms that give rise to clusters of properties that occur together with significant regularity. These mechanisms may take the form of causal relations between properties in a cluster, which favour their co-instantiation, or underlying processes that produce the same result, or both. Boyd argues, for instance, that biological species are homeostatic property cluster kinds.[3] The idea of homeostatic clustering more generally is certainly attractive, but sociability will not always be analysable in this way. For example, it would seem that homeostatic mechanisms are not responsible for the co-instantiation of the mass, charge, and spin of electrons. In the case of many essence kinds, sociability is a brute fact, admitting of no causal decomposition. The presence of homeostatic mechanisms is not the *sine qua non* of kindhood. It is a special case of sociability.

I have given an account of the kind concept that provides the semirealist with a fitting diagnosis of why different categories of objects can be expected to conform to law-like generalizations. In some cases these generalizations are strict, as is often expected in scientific disciplines whose subject matters are sufficiently fundamental or uncomplicated. In other cases and wherever *ceteris paribus* laws are found, law-like generalizations are less strict and admit of varying degrees of exceptions. In all cases the behaviours of members of kinds are governed by causal laws, which in the context of semirealism, as the reader will recall, are relations between causal properties, or concrete structures. Law statements describing these relations are causal law statements precisely because they summarize information about how altering some property instances can affect others in a causal process. Concrete structures and sociability underwrite the inductive success associated with natural kinds in the sciences.

In accounting for the success of epistemic practices involving kinds, it is important to appreciate that in many cases, some of the causal properties relevant to this success will receive no mention at all in the law-like generalizations one cites. Causal law statements employ terms or variables that

---

[3] Ereshefsky and Matthen 2005 contend that Boyd's homeostatic cluster proposal does not give an adequate account of stable polymorphism within biological taxa. An adequate account, they argue, would also have to incorporate mechanisms favouring heterogeneity and heterostasis.

can be interpreted straightforwardly as referring to causal properties, but behavioural and definitional generalizations usually do not. Consider some of the classic philosophical examples of law statements, often attributed to Carl Hempel, such as 'Robins lay bluish-green eggs' (a behavioural generalization) and 'All ravens are black' (a definitional generalization). These statements describe putative natural kinds, and it might be tempting to view them as supported by causal laws involving dispositions conferred by properties such as robinhood and ravenhood, to lay bluish-green eggs and to be black, respectively. But this, I will suggest, is the wrong way to interpret these generalizations, and the moral is that kinds can play important roles in supporting inductive practices even in the absence of a knowledge of the relevant causal laws. Let us see why this is so.

A knowledge of kinds may be useful in facilitating generalizations and predictions, but this is not always because the law-like generalizations under consideration describe a causal law. Behavioural and definitional generalizations do not describe relations between causal properties, at least not directly. If one interprets the relevant terms employed by these statements as directly referring to properties, one does not end up with descriptions of causal laws as explicated in Chapter 5. For example, the property of being a raven – granting for the moment that there is such a property – and the property of being black are often co-present, certainly. This is because raven feathers usually have a pigment that one describes as black. Being a raven, however, does not itself confer a disposition to be black. It is not part of the kind essence of a raven that it be black, or even that it have a disposition for this colour. Ravens do not have essences; they are a cluster kind. Being black is one of a cluster of properties that ravens usually do have, but need not. Since it is not necessary for a raven to be black (albino ravens are still ravens), even dispositionally, there is no relation between properties here that one could describe as a law.

According to what is often called the 'sparse' view of properties, not every predicate names a distinct property. Sometimes different predicates refer to the same property, and some predicates are elliptical for more detailed descriptions of combinations of properties. Anyone attracted by a sparse view of properties will be sceptical of properties of kindhood, like that of 'being a raven' or 'being a neutrino', simply on grounds of parsimony. The property of being a member of a kind explains nothing that is not already accounted for in terms of quantitative and determinate causal properties. Profligacy would not add to the varieties of kinds one may investigate, for example. A raven is a raven, not because it has the property of being a raven, but because it has other properties that allow one to group it together with other token birds, and the same can be said *mutatis mutandis* regarding essence kinds such as neutrinos. I am sympathetic to this view, but even without it I think one can see that the prediction 'The next raven I observe is very likely to be black' is not supported by a causal law relating anything like ravenhood and blackness. It *is*, however, supported by the fact that most organisms falling into the cluster kind raven are black, and this follows from numerous causal laws concerning the development of pigment in the feathers of most birds grouped together in this category. In such cases, connections between behavioural or definitional generalizations and causal laws may not always be transparent, but they are there nonetheless.

When it comes to making generalizations and predictions about kinds, degrees of success vary according to the extent to which their members share the same intrinsic properties. The study of sociable properties and their relations distinguishes statements of causal laws from mere but nonetheless helpful law-like generalizations. The fact that one must often take generalizations about kinds and inductive projections as helpful guides rather than absolute decrees is hardly cause for alarm. In the case of cluster kinds it is obvious why there are no decrees, since, lacking essences, their members need not possess any one of the properties associated with the set defining the relevant kind. Here, both behavioural and definitional generalizations will likely admit of exceptions. In the case of essence kinds, definitional generalizations describing members in terms of their essential properties will be exceptionless, but the same cannot be said of behavioural generalizations. Essences are no guarantee of uniform behaviour among the members of a kind, even in exactly similar circumstances. One reason for this is that some causal properties are irreducibly probabilistic, such as the disposition of a radioactive atom to decay within a given period of time. Another reason is that the other, non-essential properties of members of essence kinds are generally causally efficacious too, and as a result, many if not most laws hold only *ceteris paribus*. Causal laws, after all, typically involve dispositions that may or may not be manifested depending on the circumstances.

Physics and chemistry, the producers of paradigmatic essence kinds, illustrate this well. For example, atoms of specific elements share atomic numbers (the numbers of protons in their nuclei) as essences, but different ions of atoms (ones with different numbers of electrons) and different isotopes (ones with different numbers of neutrons) can behave in radically different ways when placed in exactly similar test conditions. Luckily, however, the prevalence of *ceteris paribus* laws is not so worrying as to prevent successful inductive practices with respect to scientific categories of things. And what more could one want from a concept of kinds?

6.5 BEYOND OBJECTIVITY, SUBJECTIVITY, AND PROMISCUITY

I have argued that the concept of natural kinds does have a substantive role to play in the world view of scientific realism after all. Earlier I suggested that the two most important connotations of this concept are the ideas of essence and objectivity. I also suggested that the primary motivation for thinking there are such things as natural kinds stems from attempts to account for successful inductive practices in the sciences, and more specifically, success in constructing generalizations and making predictions about the members of scientific categories. Until now the discussion has focused on the prospects of views according to which kinds have essences (such as NE), the importance of supplementing essence kinds with cluster kinds, and the link between both sorts of kinds and inductive success. Having covered this ground, I believe that certain consequences regarding objectivity follow immediately. Given the preliminary conclusion that ontological support for scientific practices such as generalization and prediction can be found in the relations of causal properties and the phenomenon of sociability, it is time now to consider the notion of objectivity.

In expositions of natural kinds, the idea that a system of classification is objective is usually elucidated by saying that it respects nature's own divisions, thereby reflecting the mind-independent kind structure of the world. It is this feature of scientific classification that makes natural kinds natural. In contrast, subjective accounts of kinds are typically described as ones according to which one's classificatory schemes, scientific or otherwise, are merely useful, convenient, or of interest to humans, and thus arbitrary from a nature's-eye point of view. Such categories are unreflective of mind-independent divisions. The distinction here between objective and subjective is intended to distinguish mutually exclusive understandings of the nature of taxonomy. If kinds are objective categories, it is held, desiderata such as usefulness and convenience are strictly irrelevant to the idea of correct classification (though they may coincide accidentally). Likewise it is held that if kinds are demarcated on the basis of human interests, it is likely that one's criteria for distinguishing one kind from another are merely 'nominal', to use Locke's terminology, as opposed to anything like real essences furnished by nature, assuming there are such things.

In the context of the account of kinds I have described, however, this opposition of the objective and the subjective cannot be sustained. Indeed, on this understanding of kinds, these notions are properly conflated. According to this view, to the extent that a system of classification facilitates inductive success, kind-talk picks out perfectly objective features of the world, viz. instances of sociability. This offers no prescription, however, against utility, convenience, and interests playing a role in the demarcation of kinds. There are innumerable patterns of spatiotemporal property distribution that exist objectively in nature. In the course of investigating parts of the world, the sciences recognize some of these patterns and describe them in terms of categories of objects. Many realists do offer proscriptions, *ex cathedra*, against the role of interests in this process, but as I will suggest, there are no empirically or philosophically compelling reasons to do so. Like an exclusive commitment to essence kinds, the pursuit of "one correct taxonomy" of objects is a relic of past accounts of natural kinds, properly cast aside. Understanding this in turn has important consequences for whether the adjective 'objective' offers any helpful qualification at all in the context of scientific classification. Let us consider these claims further.

The distinction between objective and subjective accounts of kind classification is usually argued for by means of some combination of three main considerations. The first is the idea that different interests or perhaps other subjective factors could produce different and mutually incompatible taxonomies, which contradicts the view that classification should reflect the one and actual (objective) natural kind structure of reality. Objectivity is typically associated here with realism about kinds, and subjectivity with antirealism. Ellis (2001), for example, suggests that although one recognizes cluster kinds such as biological taxa for everyday and scientific purposes, these are 'of our own making' and thus not objective. Kyle Stanford (1995) holds that the contextual, subjective interests of scientists play a constitutive role in biological classification, since legitimate and independent explanatory demands require distinct species concepts, and explanatory demands are relative to historical and scientific contexts. This, he says, precludes a realism about kinds demarcated by any given species concept. Ereshefsky (1998) takes a variant route to the same conclusion. Species pluralism, he thinks, suggests an antirealism about species as kinds, not because scientific interests are contextual or subjective, but rather because there is no unifying classificatory principle shared by the several, well-warranted, scientifically sanctioned species concepts found in biology today.

There are two problematic inferences here. The first is from the role of interests to an antirealism about kinds, and a second is from the mere fact of pluralism regarding kind classification to the same conclusion. Realists about natural kinds invite this sort of reasoning, because they speak as though objectivity requires that classificatory schemes reflect *the* natural kind structure of the world. The problem here is the definite article. Given that the utility of the kind concept to realists is a function of the sociability of properties, the notion that there is *one* natural kind structure is exposed as an implausible condition of objectivity. There are presumably uncountable numbers of incompatible ways of grouping properties that are sociably distributed across the natural world. So long as each of these different taxonomic systems reflects this distribution, it is difficult to see how any of them could be considered non-objective. Mutually incompatible taxonomies are all objective so long as each picks out genuinely sociable collections of properties. Likewise, scientific interests and the contingent explanatory demands of the times may influence how the sciences structure networks of kinds, but so long as they do so on the basis of properties with instances demonstrating some pattern of sociability, the resulting kinds will reflect the nature of an objective reality. Neither any subjective factor nor pluralism precludes objectivity.

Still, the association of pluralism in classification with subjective knowledge is well entrenched in the tradition of thinking about natural kinds. As I have suggested, one reason for the widespread currency of this association is the mistaken assumption that objectivity mandates only one true taxonomy. A second reason is the similarly widespread idea that whatever one's attitude towards cluster kinds, essence kinds at least are independent of human interests, and there is indeed one true taxonomy of *them*. This putative disanalogy between cluster kinds and essence kinds is in large part responsible, I suspect, for fuelling the traditional realist emphasis on essence kinds. NE exemplifies this emphasis when it claims that only kinds with essences are objective and attempts to limit genuine kindhood to these kinds specifically. There are many possible taxonomies of biological kinds, so the story goes, but there is only one plausible taxonomy of subatomic particles.

Establishing the truth of the assertion that there is only one way to classify essence kinds correctly, however, is easier said than done. It is no argument for the view that there is only one plausible taxonomy of particles, for example, to note that one's best current theory furnishes only one taxonomy of particles. The Standard Model in particle physics, which describes the natures and behaviours of these very small constituents of matter, serves the investigations of physics extremely well. Could physics have been served otherwise? It is notoriously difficult to imagine counterfactual histories of scientific disciplines, let alone evaluate their plausibility.[4] But leaving such speculations to one side, simpler considerations suggest that utility, convenience, and interests do play a role in the classification of essence kinds no less than of cluster kinds, if perhaps less conspicuously. For example, classifying atoms according to the numbers of protons in their nuclei is especially useful in facilitating various explanatory tasks in physics and chemistry, but classifying atoms according to whether they occupy ground states or excited states of energy is less useful. Thus, it is no accident that classification in these fields respects the former groupings of causal properties but not the latter. This does not suggest that the former properties are more objective, however. Nor does it suggest anything about their relative status as markers of possible, objective kinds.

A third and final reason for the traditional distinction between objective and subjective classification can be traced to the influence of Mill. In Mill's view, a natural (objective) kind is a class whose members share substantial numbers of properties. Indeed, he suggests (1846, Part I, ch. 7, section 4) that where genuine kinds are concerned, such properties are inexhaustible. The discovery of shared properties among the members of a kind is merely the tip of the iceberg – indefinitely more lie in wait. Inspired by this picture, many view kinds that fall short of Mill's standard as deficient, and thus likely subjective in one or more of the senses I have described. Though the satisfaction of this standard may be sufficient, however, it is not necessary for scientific kindhood. The more basic essence kinds, for example, such as subatomic particles, have essences comprising relatively few properties. Furthermore, though kinds of vitamins, hormones, polymers, and dyes are subjects of scientific study, it is arguable that their members may differ significantly, having little more in common than certain functions, thus sharing relatively small numbers of functional or other relational properties (cf. Khalidi 1993, pp. 107–8). Not all scientific kinds are Mill's kinds. Therefore, since realists view the sciences as furnishing objective descriptions of the natural world (subject to the usual caveats), Mill's standard should not be regarded by them as an exclusive indicator of objectivity.

There are further reasons for doubting Mill's approach. Sharing "substantial" numbers of properties is a poor test of objectivity because there are

---

[4] Redhead 1984, p. 276, hints at one intriguing possibility in Heisenberg's proposal to treat the neutron and the proton not as two different particles, but as two states of a single particle.

no good answers to the questions of why this should matter and how many would suffice. Certainly, objects that share many intrinsic properties lend themselves to greater numbers of law-like generalizations and inductive projections than those that do not, but this sheds no light on the question of objectivity. Having an impressive range of inductive significance is suggestive of natural kindhood, but it is not a necessary condition. In a partial nod to Mill, connecting the idea of many shared properties with scientific worth, Wilkerson (1995) maintains that natural kinds are those that are amenable to 'detailed scientific investigation'. In a similar vein, Ereshefsky (1992) claims that not just any similarity relations will do; those demarcating kinds should be scientifically relevant or important. Even John Dupré (1993), who advocates what he calls a 'promiscuous realism' or liberal pluralism endorsing the legitimacy of both everyday and scientific kinds, thinks one should draw a line between genuine kinds and others anthropocentrically, recognizing as genuine only classes that have everyday or scientific significance. For example, a classification of objects into groups with mass less than 1 kg, mass greater than or equal to 1 kg but less than 2 kg, and so on, he says, would be 'thoroughly artificial: we would surely not imagine that such a classification contributed in any way to our understanding of any pre-existent features of things in the world' (p. 17).

But any kind identified on the basis of the properties of its members can be investigated scientifically, and there is no principled answer to the question of how many shared properties is enough. It is possible to make generalizations and predictions in cases where objects share just a few properties or even one property. More importantly, contributing to one's understanding has nothing to do with the question of whether a kind is objective. Through the sciences one aims to learn about the natural world, but the systems of classification that best contribute to one's under-standing of the parts one chooses to investigate do not thereby constitute the limits of what is natural. It would appear that it is important to distinguish between everyday and scientific kinds on the one hand, and natural kinds on the other. Everyday and scientific kinds recognized in the course of systematizing nature are perfectly natural, but what is natural goes well beyond what is useful, convenient, or interesting in everyday and scientific contexts. Nature is composed of distributions of property instances, only some of whose patterns of sociability we consider and investigate. There are, it would seem, unimaginably many natural kinds.

My consideration of the relevance of the concept of natural kinds to the ontology of scientific realism has produced a view that is somewhat deflationary about kinds themselves. Kind-talk simply reflects distributions of causal properties. One describes cases in which distributions are sociable enough to be useful, convenient, or interesting as instances of kinds. But no classification, so long as it is made on the basis of properties in the world, whether in terms of an essence or a cluster, is more or less objective than any other. The subjectivity of the choice, generally made to serve everyday or scientific ends, is thus independent of the question of what is natural. The fact that scientific taxonomies are successful insofar as they systematically describe distributions of instances of causal properties is grist to the mill of semirealism. This is not a realism, however, founded on outmoded concepts of kinds inherited from the great metaphysicians of ancient and medieval times, married to essences and suspect notions of objectivity. Times change, and modern scientific realists should embrace a new conception of natural kinds.

In describing a proposal for what I take to be the most promising face of realism today, I have considered the nature of the dispute between empiricists and metaphysicians, the idea of selective scepticism and its realizations in entity realism and structural realism, the notion of a causal process, the nature of causal properties and laws, and now, finally, the concept of natural kinds. Semirealism commits to certain properties, concrete structures, and kinds of particulars, both observable and unobservable. Earlier I suggested that ultimately, the internal coherence of realism as a philosophical position depends on having not only a plausible account of the unobservables described by scientific theories, but also some understanding of the most central metaphysical items one invokes in giving such an account. I have thus attempted to illuminate the foundational supports of realism and to describe at least one package according to which it is, in fact, an internally consistent and coherent stance. Some might worry that in doing this, however, I have started something the realist cannot finish. Part of the traditional empiricist critique of speculative metaphysics is the charge that it invites a regress of explanations. Equipped with an understanding of properties, structures, and particulars, is the realist now required to give an account of the metaphysical basis of *this* understanding, and so on and so forth, *ad infinitum*?

The answer, I think, is no. A scientific realist needs just enough metaphysics to support the commitments of semirealism, and I have left further questions open for contemplation elsewhere. The finer-grained ontological status of properties and particulars, for example, is a case in point. In other cases, I submit, there is nothing deeper to be explained. Consider the causal properties in which semirealism is grounded. Why do they confer some dispositions but not others? One might well respond that the fundamental natures of things admit of no further explanation. Kuhn (1977/1971) suggests that after losing out to mechanistic explanations in the seventeenth century, Aristotelian-type explanations in terms of natures began to recover in mechanics before returning with a vengeance in nineteenth-century physics and beyond. The properties in terms of which these explanations are given, such as the spins of subatomic particles, are with us today. I have described the natures of causal properties and particulars in terms of dispositions for concrete structures, and sociability. As every empiricist knows, to ask questions beyond a certain point is meaningless. The empiricist draws a line separating the explanatory from the non-explanatory between the observable and the unobservable. Realists draw their line in a different place, the proposed location of which, I hope, is now clearer.

PART III

Theory meets world

CHAPTER 7

Representing and describing: theories and models

7.1 DESCRIPTIONS AND NON-LINGUISTIC REPRESENTATIONS

The primary goal of this work has been to propose a metaphysics for scientific realism, and with much of this project now in hand we are well placed to confront one last constellation of issues. Part I focused on the question of what realism is and what it has become, tracing the evolution of the position over recent history and in response to specific forms of anti-realist scepticism. In the course of that discussion I fused what I think are the most promising features of these developments and called the resulting package 'semirealism', reflecting the graded commitment and epistemically selective attitude characteristic of sophisticated versions of realism today. In Part II, I developed a proposal for the key foundational concepts of semirealism, plausible accounts of which are important to the internal coherence of a realist approach to scientific knowledge. There I considered the nature of causal processes and causal properties, dispositions and necessity, laws of nature, and the place of natural kinds within scientific taxonomy. Equipped with this framework, I believe I am now in a position to begin the process of connecting, more explicitly, the metaphysics of semirealism with certain aspects of its epistemology.

In many philosophical contexts it is difficult to separate epistemic considerations from metaphysical ones, and this is very much the case in the broader context of scientific realism. It is, after all, a thesis or a stance concerning the interpretation of scientific knowledge. Though a reflection on the internal coherence of realism and the development of a metaphysics capable of serving this end have been my primary objectives, several topics in the borderland of the metaphysics and epistemology of realism naturally emerge from the discussions of Parts I and II. At various points earlier on, for example, I touched on the issue of how causal properties and concrete structures, which together form the bedrock of semirealism, are described by scientific theories. The notion of description is central to realism, and it is time now to consider it in more detail. In Chapter 8, I will explore the question of how scientific description can be thought to yield knowledge, given that (as realists commonly acknowledge) one's best theories are often and perhaps even typically false, strictly speaking. Before this, however, several preliminary issues regarding the modes of such description require attention. In this chapter I will consider the tools used to contain scientific knowledge, and how these tools are employed to furnish descriptions of parts of the world. In other words, I will consider the representational status of theories and models.

Perhaps the best way to begin this task is by thinking about what it means to speak of representations and descriptions in connection with scientific knowledge. These terms are often used more or less synonymously in everyday conversation, but it will prove useful to distinguish them more carefully in the scientific context. A representation is something that stands in an asymmetrical, "intentional" relation with something else. That is, it is something that is *about* another thing, and this "about-ness" is not usually reciprocated. There are some details concerning the precise characterization of what it is to be a representation that will not concern me here. For example, one might wonder whether the intentionality of a representation is a mind-independent property of it, or whether it is rather something a representation has as a consequence of being conceived this way by a thinking agent. I suspect that in most if not all cases the representational status of a thing depends on its being conceived or at least used as such, but in any case these details will play no role in the present discussion. Furthermore, though canonical examples of representations such as diagrams, drawings, paintings, and sculptures are usually observable, this should not be taken as a necessary condition of representation. Ideas, for instance, are often held to have representational content. Theories and models understood as abstract (and thus unobservable) entities are likewise representations, and I will take this for granted in what follows.

In addition to the idea of being about something else, there is a further connotation of being a representation that demands some attention here. This is the idea that a successful representation contains information regarding the thing it represents. As we shall see in Chapter 8, sometimes this information is rather minimal: in the limit, it may be exhausted by the fact that the subject of the representation exists, in which case the notion of representational information collapses into that of intentionality. In many cases, however, representations bear some substantive similarity relation or relations to the things they represent, and unlike intentionality, these relations are symmetrical. Such representations *share* something with their subject matter. What counts as a "substantive" similarity relation, however, can vary greatly depending on the representational context. What counts as substantive can again be rather minimal. A football coach can represent defenders and attackers with chalk-mark Xs and Os on a blackboard, for example, with the hope of teaching the team some tactics, even though the players themselves presumably have little in common with chalk-mark Xs and Os! The relevant similarity relations here between blackboard representations and what hopefully takes place on the pitch concern only certain spatial relations between distinct objects. But with luck there is some similarity nonetheless, however attenuated, and this leads to the issue of description.

Anything that can be regarded as a description will have the features I have just outlined for representations. That is, a description is about something, and contains information about the thing it describes. Descriptions are a proper subset of representations, and thus it is not surprising that 'representation' and 'description' are often used interchangeably in the context of everyday conversation. There is at least one feature of descriptions, however, that is not shared by all representations, and it turns out that this feature is important to discussions of scientific knowledge. Descriptions are essentially linguistic, and this attribute distinguishes them from other sorts of representation. Descriptions are representations that are composed of pieces of language, construed broadly to include expressions in natural languages (such as English), expressions in logical languages (such as the first-order predicate calculus), and mathematical expressions. Many representations including several of the types mentioned a moment ago are not linguistic entities in any of these senses. Whenever instances of language are used to represent something, these instances belong to the subset of representations one calls descriptions.

One might worry that the contrast I have sketched between descriptions and other representations is too stark. Is it not reasonable to think that non-linguistic representations describe the things they represent? Surely a diagram, for instance, describes its subject matter? No doubt this is so in a loose manner of speaking, but some care is required here in understanding the employment of these familiar terms. Any representation can be *used* to describe the thing it represents, but this does not imply that all representations are descriptions. One may describe a non-linguistic representation in order to point out similarity relations between it and the thing it represents, or the absence thereof. That is, one may formulate

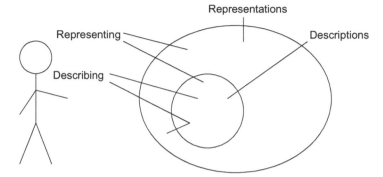

Figure 7.1. Using representations and descriptions

descriptions in a language in order to reveal the information contained in a non-linguistic representation about its subject matter. For example, one may describe a portrait, itself a non-linguistic representation, to the end of considering whether it constitutes an accurate or otherwise pleasing representation of the thing it portrays. But this does not suggest that non-linguistic representations are descriptions. Rather, it suggests that non-linguistic representations are capable of being described. The sciences undertake to represent but also to describe, and though they often use non-linguistic entities to represent aspects of the world, they generally employ descriptions in order to cash out the knowledge they contain. As we shall see, this point about representation and description is crucial (so much so that I have represented it in Figure 7.1) to understanding the relation of theories and models to the world. I will return to it shortly.

With this opening sketch of the distinction between representation and description in mind, let me now introduce the other major focus of this chapter, the putative distinction between theories and models. The relationship of theories and models and the question of whether they are identical or distinct, more specifically, has been a controversial matter for several decades. I believe, however, that this debate is largely orthogonal to the issue of what realism is and how it is best understood. Some proposals for how to construe the nature of scientific theories do have weaknesses, and anyone interested in this question does well to avoid them. Granting this, however, it would seem that issues surrounding the nature of theories are largely matters of convention for the realist, with no important epistemic implications. Many realists do not agree with this assertion, however. Some contend that taking a specific stand on the nature of theories is advantageous for realism. In the rest of this chapter I will aim to show why

I think this contention is mistaken, as a means to the end of clarifying what is actually at stake for the realist here. The debate concerning what a theory is, it turns out, is a red herring so far as realism is concerned, but an explanation of why and how this is so will help to illuminate important connections between knowledge, representation, and description in the sciences.

The issue of the nature of theories is a tempting distraction not only for realists, who endorse knowledge claims regarding both observables and unobservables, but also for many empiricists and other antirealists who limit knowledge claims to the observable. Though I will continue to focus on the commitments of realism here, it should be noted that one consequence of the arguments of this chapter is that the "ontological" status of theories is largely insignificant, epistemically speaking, for *anyone* who interprets them as containing knowledge of the world, however that knowledge is characterized. The nature of theories can be construed in different ways; their epistemic significance stems from the fact that however they are construed, they can be used to describe the world. The nature of theories apart from this epistemic function is largely and merely a matter of convention. In order to see this, it will help to have an understanding of what the options for a choice among conventions might be. Let us consider these possibilities in broad outline now.

7.2 REPRESENTING VIA ABSTRACTION AND IDEALIZATION

Theories are repositories for scientific knowledge. They embody our most considered beliefs about the nature of the world. Beyond this very basic characterization, however, views concerning the nature of theories diverge considerably. The traditional view is that a theory is a linguistic entity, such as a collection of statements, that purports to describe some aspect of the world. Within this broad tradition different conceptions are possible, the most famous of which is inherited from the work of the logical positivists and logical empiricists. On this proposal theories are axiomatic systems of statements closed under deduction (systems containing axioms plus any statements deducible from them). Emphasizing the role of logic in systematizing everything from scientific knowledge to practices such as theory confirmation and explanation, logical empiricists popularized the notion that theories conceived this way can be expressed in a formal language, such as first-order logic, whose elements are characterized by a specific syntactical structure. The emphasis here on the syntax of linguistic formulations led critics to dub this the *syntactic* view of theories, and the implausibility of this proposal as a general account of theories formed part of the body of criticism that ultimately resulted in the rejection of logical empiricism in the latter part of the twentieth century.

I will not digress here to consider the difficulties associated with the logical empiricist conception of theories (see Brown 1977, chs. 1–3; van Fraassen 1980, ch. 3). Let it suffice to say that tying the concept of a theory too closely to the idea of a linguistic formulation engenders several challenges. On such a view it is difficult to explain, for example, the apparent fact that one and the same theory can be expressed by means of different linguistic formulations, whether in terms of natural languages or in terms of formal ones, as in the case of classical particle mechanics, which can be given a Hamiltonian or a Lagrangian formulation. It is important to appreciate, however, that the traditional understanding of theories as linguistic entities need not be construed in the manner proposed by the logical empiricists. A degree of independence from formulations in language can be achieved by identifying theories, not with such formulations themselves, but with the propositions expressed by them. Focusing on propositions or the content of statements as opposed to statements themselves has certain advantages. For one thing, it suggests a construal of theories on which they are in principle expressible in different ways. This should not be taken to indicate that the propositional view is free of challenges of its own, however. The very existence of propositions as abstract entities independent of linguistic formulations is contestable, and even granting an ontology of propositions, it is unclear to what extent they can or should be understood as syntactically neutral.[1]

The main alternative to the possibility of construing theories in linguistic terms arose in response to difficulties associated with the syntactic view. A number of closely related accounts have coalesced around this alternative approach, which is commonly referred to as the *semantic* view of theories. The term 'semantic' here is used in the sense of formal semantics or model theory in mathematical logic. As Suppe (1989, p. 4) neatly summarizes, the semantic view 'construes theories as what their formulations refer to when the formulations are given a (formal) *semantic* interpretation'. On this approach theories are not linguistic but rather abstract, set-theoretic entities. They are models of their linguistic formulations. According to the semantic view, a theory is a family of models, where a model is any object that satisfies the linguistic formulations commonly

---

[1] For a recent discussion of propositions incorporating elements of syntax, see King 1995. Cf. Niiniluoto 1998 on Carnap's Q-predicates.

associated with the theory. For this reason the position is also called the *model-theoretic* or *model* view of theories. The precise description of these extra-linguistic entities varies according to different versions of the position. Some prefer to think of them in terms of set-theoretic predicates, others in terms of state spaces, and Suppe himself prefers what he calls 'relational systems' (see his Prologue for a summary of these accounts). These particular details are inconsequential for present purposes, however, so I will simply use the generic term 'model' in speaking about the semantic view henceforth.

Another difference between versions of the semantic view concerns the extent to which models are conceived as separate from the linguistic formulations that can be used to describe them. For some proponents, a model is not merely something that satisfies certain axioms but also something that includes a mapping from some elements of a linguistic formulation to some elements of the model. This sort of account is favoured by Giere (1988, pp. 47–8), who suggests that a model can include, for example, a function that assigns subsets of objects to one-place predicates, two-place relations, and so on. Others hold that there should be a stronger separation between models and expressions in language, stressing that the relation between a linguistic formulation and its models is purely one of definition. On this version, models are by definition merely those things that satisfy, for example, the mathematical equations associated with a theory. This sort of account is favoured by French and Ladyman (1999, pp. 114–18), according to whom theories should be understood as abstract entities that exclude the linguistic formulations defining them. This difference of opinion among supporters of the semantic view essentially concerns the issue of how best to achieve an appropriate level of independence for theories from language.

It is not my intention here to resolve the finer points of either the linguistic view or the semantic view of theories. My present interest is in the question of whether one's view of the nature of theories has any bearing on the issue of realism, and indeed, at least one connection has been suggested. Some advocates of the semantic view think that this approach to theories helps to facilitate realism, and this thesis merits consideration. Though I believe it to be mistaken, an exposition of the cases for and against it will shed further light on the semirealist account of scientific knowledge. The primary motivation for the semantic approach is to avoid concerns about the relation between linguistic entities and the world. In addition to the worries mentioned a moment ago regarding the syntactic view, several more general challenges arise from the philosophy of language. One of the most important of these for realism is the challenge of determining whether the nature of the relation between elements of language and aspects of the world should be conceived in terms of some sort of correspondence. It may seem, *prima facie*, that if one were to adopt a view of theories according to which they are objects as opposed to pieces of language, this challenge would evaporate. Difficulties inherent in the attempt to forge links between language and the world, one might think, simply vanish if one is rather focused on forging links between non-linguistic entities. Perhaps models are better suited to representing the world than descriptions are.

On the semantic view, theories are families of models, and to begin my investigation into the thesis that adopting some version of it is advantageous to realism, it will help to consider briefly how these models represent. Recall that a representation is about something and contains information about that something. As representations, models naturally invite comparisons to the things they model, and in scientific contexts models often contain rather caricatured information about the things they represent. For example, consider the simple pendulum. This model correctly describes the motion of a mass attached to a frictionless pivot by means of a massless string, swinging in a uniform gravitational field and encountering no resistance. That models are often caricatures of reality is of course widely appreciated, but what is perhaps less well appreciated is that there are two quite different processes involved in their construction. I have mentioned both of these processes earlier, in the context of *ceteris paribus* law statements and vacuous laws in Chapter 5. One of them I called 'abstraction', and the other I called 'idealization'. The distinction between abstraction and idealization will prove an important tool in the discussion to follow concerning the import of the semantic view for realism. Let us consider this distinction in more detail now.[2]

Abstraction is a process in which only some of the potentially many relevant factors present in reality are represented in a model or description concerned with some aspect of the world, such as the nature or behaviour of a specific object or process. Here one excludes other factors that are potentially relevant to the phenomena under consideration. Abstraction is a common feature of model-building in the sciences for at least two reasons. Firstly, the potentially relevant factors with respect to a given class

---

[2] Suppe 1989, pp. 82–3, 94–9, uses these terms in a very similar way. McMullin's (1985) distinction between 'causal idealization' and 'construct idealization' mirrors the one here between abstraction and idealization to a great extent. Cartwright's (1983; 1989, ch. 5) characterization of abstraction and idealization is also similar, though considered with respect to a set of issues largely different from the ones discussed here. For a detailed and very systematic treatment, see Jones 2005.

of phenomena are often very numerous, making the construction of an equally refined model somewhat if not highly impractical. Secondly, the relative importance of many and sometimes most potentially relevant factors is often negligible given the explanatory purposes and required levels of predictive accuracy relevant to specific contexts of theorizing and experimentation. Pragmatic constraints such as these play a role in shaping how scientific investigations are conducted, and together determine which and how many potentially relevant factors are incorporated into models and descriptions during the process of abstraction. The role of pragmatic constraints, however, does not undermine the idea that putative representations of factors composing abstract models can be thought to have counterparts in the world. The fact that some factors are ignored is perfectly consistent with the idea that others are represented.

Idealization is also commonplace in model construction, and here too, pragmatic constraints enter the picture. The trademark feature of idealization is that model elements are assembled in such a way as to differ from the things they represent, not merely by excluding factors as in the case of abstraction, but by incorporating factors that cannot exist as represented given the actual properties and relations involved. 'Cannot' here can be understood in terms of the account of *de re* necessity and possibility outlined in Chapter 5. For example, models in classical mechanics generally represent the masses of objects as though they are concentrated at extensionless points, but this is an idealization. One does not actually think that mass is concentrated this way, nor does one think that such a thing is possible in worlds where objects with mass exist, given the nature of mass properties and the dispositions they confer. Abstraction involves choosing some factors and excluding others, such as air resistance in the model of the simple pendulum, but idealization involves simplifying the natures of factors that have been chosen. These are not mutually exclusive categories, of course. Indeed, many representations are both abstract and idealized; earlier I used the term 'pure abstraction' to label cases of the former that are not also the latter. Models and descriptions in the sciences and more generally are almost always abstracted and idealized representations of aspects of the world.

The distinction between abstraction and idealization would seem to have important implications for a realist interpretation of scientific knowledge. At first glance it appears that semirealists should adopt a significantly different epistemic attitude towards these practices, since pure abstractions are (or can be associated with) correct descriptions of causal properties and relations manifested in at least some circumstances. On the other hand, idealizations are (or can be associated with) descriptions of properties and relations that do not and cannot exist as described in any circumstances. An assessment of the significance of the difference between abstraction and idealization, however, is complicated by the connections between them. In a specific context, for example, one might use an abstraction in the manner of an idealization, by applying it to circumstances in which the relations described could not apply, given the natures of the properties involved. In circumstances where factors not represented are present, the dispositions manifested may differ from those described by an abstract model. I will consider these and related issues in Chapter 8. In the meantime, we now have enough of a sketch of abstraction and idealization to assist in a consideration of whether focusing on models can help to facilitate a plausible realism. Let us return now to this suggestion on behalf of the semantic or model view of theories.

7.3 EXTRACTING INFORMATION FROM MODELS

The semantic approach may well furnish an attractive account of what scientific theories are. It does not, however, offer any sort of shortcut or special aid to realists. As I mentioned earlier, this latter assertion runs contrary to the intuitions of some advocates of the model view, and it is time now to consider these intuitions in more detail. I will attempt to show that they are mistaken by presenting the realist with a dilemma. Realism inescapably involves making substantive claims about the world, regarding both its observable and its unobservable aspects. Thus, realists must either make such claims or give up their realism. The dilemma I have in mind stems from the idea that substantive claims about the world are descriptions, as I have elucidated them, and descriptions are essentially linguistic. Non-linguistic representations do not by themselves constitute the sorts of claims that are required in order to express a realist commitment. In order to express commitments, non-linguistic representations must be described, and this leads to a dilemma for realists as follows. They can either invoke descriptions and thereby face up to traditional challenges associated with the interpretation of language, or they can abandon any substantive realist commitment. One cannot be a realist *and* dodge the challenges associated with interpreting language, and therefore, the semantic view offers no special facilitation of realism.

As it turns out, versions of the argument just outlined in summary extend well beyond a consideration of realism. If the core idea of the argument (regarding the importance of descriptions for knowledge) is compelling, then no one who believes that theories yield information about the world, including empiricists who believe only the observable consequences of theories, will find any special facilitation of their positions on the semantic approach. Indeed, advocates of this approach generally do not contend that it favours any specific epistemology of the sciences whether realist, empiricist, or anything else (cf. van Fraassen 1985, p. 289). They do claim, however, that because of its emphasis on models as opposed to language, the semantic approach gives an account of theories that results in a less problematic treatment of scientific knowledge than linguistic accounts of theories. It is this weaker claim that is the wider target of my dilemma, and I will limit the discussion here specifically to the context of realism. The moment model theorists make any sort of commitment, realist or otherwise, they open a door to the very difficulties that some hope the semantic approach leaves behind, such as the idea of correspondence between language and world. An understanding of the information a theory contains regarding the ontology of a given particular or process, for example, can be had only via descriptions, and not via non-linguistic representations alone.

What reason might a proponent of the semantic view have for thinking otherwise? Perhaps the most obvious difference between descriptions and non-linguistic representations provides a reason. Surely it is a simpler matter, one might think, to evaluate the information contained within theories if they are models as opposed to linguistic things. For surely models are ontologically more similar to the parts of the world they represent than are linguistic entities. If theories are models, one may compare like with like, which is easier presumably than giving an account of the relation between linguistic entities and the world. It is easier to compare two non-linguistic entities than it is to understand the relation between a linguistic and a non-linguistic entity. But this, I submit, is misleading, for it is unclear what 'easier' could mean here. One might think that a model is more easily compared to the world because both can be visualized, and visualization aids comparison. But this is to appeal to a purely metaphorical sense of 'visualization'. Much of the world of interest to realists, recall, is unobservable, and many models, especially in the sciences, are abstract objects as opposed to things that can be detected let alone observed. In most cases it is thus unlikely that comparing models to the world is any more transparent a task than interpreting descriptions of models and the world. Indeed, in most cases there seems no other way of making comparisons than to describe the relevant models and *thereby* compare them to whatever they represent.

Even in cases where models are not abstract objects at all, but rather concrete, observable things (for example, Watson and Crick's demonstration model of the structure of DNA molecules), the information that non-linguistic representations putatively contain regarding aspects of the world cannot ordinarily be accessed without descriptions. Theories do not function merely to replicate or imitate the world; they are also supposed to tell one something substantive about the things they represent. That is the point of constructing them! Barring rare exceptions, a model can tell one about the nature of reality only if one describes some aspect or aspects of it, and goes on to assert that these aspects have counterparts in reality that are similar to them in specified ways. In other words, in order to be a realist, some sort of explicit assertion of correspondence between a description of some aspect of a model and the world is generally unavoidable. There is no other way to express a realist commitment than to employ linguistic formulations to this end, and interpreting these formulations in such a way as to understand what models are telling one about the world is the unavoidable cost of realism. Generally, theories do not tell one anything substantive about the things they represent unless they employ a language.

Perhaps this is unfair to the model theorist. After all, none would contest the fact that theories can be given linguistic formulations. What the proponent of the semantic view denies is that theories should be *identified* with these formulations. Recall, however, that one of the primary motivations for this denial is to escape the perceived difficulty of having to deal in the currency of language, and this does seem at odds with the contention that descriptions are crucial to assessments of scientific knowledge. Regardless of whether one holds that theories themselves should be understood in linguistic or non-linguistic terms, the notion of independence from language cannot be sustained when it comes to thinking about how theories are used. Thus, the "ontological" nature of theories, as disputed by the traditional linguistic conception and the semantic approach, is irrelevant to the issue of how information about the world contained within theories is expressed and ultimately accepted or rejected as scientific knowledge. *Using* theories involves formulating and interpreting descriptions, and this point is independent of the question of what a theory is, precisely.

Given the central role of descriptions in considering scientific knowledge, some may be tempted to make a stronger point here about the tenability of most versions of the semantic view. If theories are strictly separate from their linguistic formulations and if linguistic entities are required in order to express epistemic commitments, it would seem that theories themselves are in principle incapable of being true or false, since the predicates 'true' and 'false' are properly applied only to constructions in language, not to objects *per se*. On the model view, one could say that although theories are neither true nor false, strictly speaking, descriptions of them have truth values when applied to specified aspects of the world. But then, in order to carry out the epistemic functions ordinarily associated with theories, one must introduce extra-theoretical devices (descriptions), and some will find this consequence strange. Whatever theories are, should they not be things that can be described as true or false (or approximately true)? If not, theories themselves might appear to have the same epistemic status as metaphors and analogies – at best good or bad, but not true or false. Realists, empiricists, and others, however, make precise commitments regarding scientific claims and have views concerning which of them constitute knowledge of the world. Are these interpreters of scientific knowledge well served by an account of theories according to which they are akin to metaphors or analogies?

I believe that this worry overstates the predicament faced by advocates of the semantic view. It is a useful objection nonetheless, however, for it illuminates the fact that one's choice regarding the nature of theories may have consequences for how one characterizes one's epistemic commitments. Though on most versions of the semantic view it would amount to a category mistake to say that theories are true or false, this hardly matters so long as descriptions of them can be used to express knowledge of the world. The semantic view furnishes no impediment to scientific knowledge, let alone epistemological positions such as realism and empiricism, and to insist that theories themselves must have truth values is simply to beg the question against the semantic view in favour of a linguistic conception of theories. This raises a more important point, however. What has been gained in the shuffle, from descriptions to models and back again? It would appear that with respect to the issue of linguistic independence nothing has been gained, since even on the semantic view, knowledge of the world depends on formulating and interpreting descriptions in connection with the things they represent. Thus, an emphasis on models does nothing to eliminate the currency of language when it comes to the project of determining whether and how aspects of theories can be cashed out as descriptions of the world. This project is the central concern of realism.

In the remainder of this chapter I will examine specific accounts of the semantic approach to theories offered by realists who believe that focusing on models can help to facilitate a realist epistemology. Considering some examples will, I hope, put some flesh on the bones of the preceding discussion regarding representations and descriptions. As we shall see in the cases to follow, contrary to the suggestion of several authors, an appeal to models does not provide a plausible realism with any more support than it would have otherwise. Perhaps the best place to begin these illustrations is with Suppe's version of the semantic view, for his account exemplifies a modest proposal for the usefulness of models to realism. According to Suppe (1989, p. 90), a model is a 'physical system', which he defines as 'a relational system consisting of a domain of states and a sequence defined over that domain; the sequence is the behaviour of the physical system'. A given physical system, he says, 'may be construed as the restriction of the theory to a single sequence'. The state of a physical system at a time is defined as the set of simultaneous values of its parameters. The behaviour of a system is its change in state over time as governed by laws, which are conceived as 'relations which determine possible sequences of state occurrences over time that a system within the law's intended scope may assume' (1989, p. 155).

Suppe is well aware of the central importance for epistemology of giving some account of the relation between aspects of models and those parts of the world they ostensibly represent. This moves him (1989, pp. 422–3) to say that

abstract structures ... do not become scientific theories until they are provided with physical interpretations (mapping relations between theory structure and phenomena). Further, it is clear that these physical interpretations are not explicitly stated ... but are implicitly or intensionally specified and are liable to alteration, modification, or expansion as a science progresses.

Characterizing the relation between models and the world ('physical interpretations') as 'implicitly or intensionally specified' is an interesting but puzzling move. 'Intensionality' here is to be distinguished from what I earlier called the 'intentionality' or the "about-ness" of a representation. The intension of a term is variously described as its meaning, or the set of connotations associated with it, or the things one must know in order to identify its extension, the set of things to which the term refers. Intension is a concept that is most commonly applied to linguistic entities, and though there are different views about how best to understand the intensions of things like words, the possibilities are relatively clear as conceived within the philosophy of language. The situation is very different, however, in the context of non-linguistic representations. It is far from clear whether the concept of intension applies here generally if at all, let alone how it might best be understood. Perhaps some metaphorical sense can be made of the idea that objects have intensional content, but this sense is not so clear as to furnish a basis for a realist account of the information models contain regarding aspects of the world.

And even if a clear account of the intensions of non-linguistic representations were forthcoming, this would not facilitate the independence from language some hope the semantic view permits. In the epistemic context in which theories are used, even if *ex hypothesi* the intensions of models determine their physical interpretations, the question of how one knows the intensions of these representations is paramount. One cannot determine the intension of a model merely by somehow contemplating its implicit qualities. Assuming there is such a thing as the intension of a non-linguistic representation, the only means by which it can be determined unambiguously is by explicitly describing some features of it and then considering these descriptions in connection with parts of the world. For there is no *one* interpretation of a model. Realists and empiricists, for example, usually interpret the same scientific models in very different ways, owing to their contrary epistemic commitments, and these divergent interpretations are revealed only by means of explicit descriptions, not "implicit specifications".

A moment ago I said that Suppe's version of the semantic view makes a modest claim with respect to the issue of support for realism. Let us turn our attention to the realism part of this equation now. Signalling the attenuated nature of the position he favours, Suppe calls his view 'quasi-realism'. Like most realists, the quasi-realist holds that some scientific theories or parts of theories can be understood as true (or something close by) in connection with both observable and unobservable aspects of reality. The quasi-realist, however, evaluates the truth of a theory in a very specific way. On this account, truth is understood in a counterfactual sense. A true theory characterizes aspects of the world counterfactually insofar as it describes what the nature or behaviour of these aspects would be like if they were the way the model describes them. If this were all quasi-realists had to say about truth, however, their modest proposal would be rather too modest – for realism, at any rate. Given this counterfactual sense of truth alone, *any* theory counts as true, no matter how outlandish or divorced from reality, since every theory describes what the world would be like if it were, in fact, as the theory says it is! In order to rescue this account from triviality, the notion of counterfactual truth must be supplemented by some other condition on what it means for a theory to be true.

Suppe does not of course believe that all theories are true. Recognizing the need for some further restriction on the notion of truth for theories, he (1989, p. 67) invokes a distinction between what he calls 'logically possible' and 'causally possible' systems (models). Any theory, he says, describes a class of logically possible systems, but if a theory is true it describes a class of causally possible systems. The set of causally possible systems is a proper subset of logically possible systems whose members do or could correctly describe things in the world in situations where factors not represented in these models exert a negligible influence. The modality introduced here by claiming that causally possible systems 'could' describe the world, if the possibilities they describe come to pass, seems to imply some sort of natural necessity. Though Suppe does not discuss necessity in connection with quasi-realism, for present purposes one may understand it in just the manner proposed in Chapter 5. That is, let us understand the things represented by Suppe's logically possible systems in terms of what I earlier described as the strongest form of necessity and possibility, and the things represented by his causally possible systems in terms of the proposal I described for necessity and possibility congenial to semirealism.

Thus combining the idea of a counterfactual condition with the idea of causal possibility, the quasi-realist arrives at a more promising definition of truth for theories. Consider a true theory incorporating a model, S. 'If P were an isolated phenomenal system in which all other parameters exerted a negligible influence, then the physical quantities characteristic of those parameters abstracted from P would be identical with those values characteristic of the state at t of the physical system S corresponding to P' (Suppe 1989, p. 95). Note that this characterization of a model, S, conforms precisely to what I previously defined as abstraction. In cases of pure abstraction – that is, cases of abstraction excluding idealization – the quasi-realist notion of truth does appear to be compatible with scientific realism more generally. For in cases of pure abstraction, factors built into models correspond to the very sorts of causal properties, structures, and particulars in the world described by the models in question. As I have noted on a few occasions now, the fact that there may be other factors that are potentially relevant to the phenomena at issue does not preclude the possibility of realism regarding those represented.

The case of idealization, however, is not so straightforward. Idealized models are constructed in such a way as to incorporate features that contradict our beliefs about some of the aspects of the world they putatively represent. In these cases there are no circumstances in which the relevant systems in the world are correctly described, because idealized theories describe things that do not and cannot exist. And while increasing the number of factors built into models may lessen the degree of abstraction, and thereby no doubt help to facilitate greater accuracy of prediction, this by itself will not make idealizations true. Idealization is commonplace in the sciences, but there is no obvious place for it on Suppe's versions of realism and the semantic view. According to quasi-realism, true theories correctly describe causally possible systems, but the theories that realists accept often describe causally *im*possible systems. This conundrum suggests a broad challenge for sophisticated versions of realism generally, and I will return to it in Chapter 8. In the meantime, let us move on to consider another proposal concerning the truth of theories and the role of models in facilitating realism on the semantic approach.

7.4 THE INESCAPABILITY OF CORRESPONDENCE

Like all proponents of the model view, Giere advocates a degree of independence from language where theories are concerned. As model theorists go, however, he inhabits the more liberal end of the spectrum by allowing linguistic entities a constitutive role in theories. On his (1988, p. 85) version of the semantic view, a scientific theory is made up of 'two elements: (1) a population of models, and (2) various hypotheses linking those models with systems in the real world'. Giere understands hypotheses as linguistic entities (more specifically, propositions) used to assert a relationship between a model and some aspect of the world. A hypothesis is true or false depending on whether or not the relation it asserts obtains. The relation asserted by a hypothesis is one of similarity of model elements to parts of the world, specified in terms of relevant respects and degrees. Consider as an example the following hypothesis (1988, p. 81):

The positions and velocities of the earth and moon in the earth-moon system are very close to those of [i.e. represented by] a two-particle Newtonian model with an inverse square central force. The earth and moon form, to a high degree of approximation, a two-particle Newtonian gravitational system.

Giere holds that his version of the semantic view supports what he calls a 'constructive realism', and more recently has extended this account into a position he calls 'perspectival realism'. His views regarding the use of models and realism, however, are vulnerable to the worries I suggested previously for realists in the context of representation and description. Given that realism involves making substantive claims about aspects of the world, realists must either make such claims or abandon realism. Substantive claims, however, are descriptions, and descriptions are linguistic entities. Thus, if realists are to remain faithful to their epistemic commitments, they must be prepared to grapple with traditional challenges associated with the interpretation of language. The model view of theories provides no escape route with which to avoid them. Giere's most ardent hope in this regard is that by emphasizing models, the realist can avoid having to make sense of the idea of correspondence between language and the world. As we shall see, however, this hope is susceptible to my dilemma for realists on the semantic view.

The core of the issue concerning whether models facilitate realism is bound up with the matter of linguistic independence, so let me begin by focusing on this. According to Giere, one of the main benefits of adopting the model view is that by doing so, realists are rescued from the need to posit a direct relationship between language and the world. Such relations, he claims, are rather indirect, via the intermediary of theoretical models. Much like the advice of quasi-realism, however, this suggestion by the constructive realist may appear somewhat puzzling at first glance, for one may wonder why a distinction between the directness and indirectness of a relation *involving language* should matter to a consideration of linguistic *independence*. More importantly, no matter how many models one stacks between linguistic formulations of theories and the world, language unavoidably enters the picture very directly, in the form of hypotheses, as soon as one attempts to determine what information theories contain about the things they represent. Making such determinations is something that realists (and most empiricists and antirealists) must do. The fact that models stand between the linguistic formulations defining them and aspects of the world cannot establish the ideal of linguistic independence, given that hypotheses stand between models and reality. Hypotheses are required to describe how elements of models are similar to aspects of the world, so there can be no linguistic independence here.

Constructive realism has a response to this objection, stemming from Giere's contention that the model view saves the realist from having to deal with issues of correspondence. Theoretical hypotheses expressing similarity relationships, he says (1988, p. 82), 'are indeed linguistic entities ... But for these a "redundancy theory" of truth is all that is required', as opposed to a correspondence theory. Realism conceived on a linguistic account of theories makes the mistake of attempting to 'forge a direct semantic link' between statements defining models and the world, but the model theorist avoids this error by maintaining that models represent parts of the world in virtue of similarity relationships, not relationships between linguistic entities and reality. This attempt to substitute relations of similarity for relations of correspondence, however, is unsuccessful. For one thing, though Giere conflates them, it is important here to distinguish the general issue of correspondence between language and the world from the more specific idea of a correspondence theory of truth. Secondly, it turns out that even if one were to favour a redundancy theory of truth in connection with hypotheses, this would not dissolve the issues of correspondence a realist must face. Let us consider these points in some detail.

Recall the example of the hypothesis regarding the Earth and moon described a moment ago: the Earth-moon system is similar to a two-particle Newtonian gravitational system with respect to positions and velocities within some satisfactory margin of error. If all one means by 'similar' here is that the relevant Newtonian model generates values for positions and velocities that match the values generated by our detections of these properties reasonably well, perhaps a redundancy theory of truth is all that one requires. The redundancy theory states that 'It is true that p' can be rephrased without any loss of meaning or semantic content simply as 'p', where 'p' stands for a statement or proposition. On this view, 'it is true that' is redundant so far as semantics is concerned, though the truth predicate may have other, pragmatic functions (adding emphasis, for example) depending on the context of its use. Thus, on the redundancy theory, to say that it is true that the Earth-moon system is similar to a two-particle Newtonian gravitational system with respect to positions and velocities to a specified degree is merely to say that the Earth-moon system is similar to the Newtonian model in those respects and to that degree.

There are several versions of the correspondence theory of truth, but all hold that 'It is true that p' indicates that some sort of correspondence relation obtains between 'p' and the world. Explicating the nature of this relation, however, is not easy, and different versions of the correspondence theory give different accounts of it. Advocates differ, for example, as to whether 'truth' should be understood as naming a specific property shared by all true propositions. In any case, the correspondence theory of truth is not a requirement of realism. Indeed, different realists hold different views regarding the nature of truth, and different views are compatible with realism. The idea of correspondence more generally, however, is crucial here, since according to realism, what determines whether a claim about reality is true are the natures and behaviours of things in the world, or as they are sometimes called, "truthmakers". This commitment can be expressed in a number of different ways: by adopting a correspondence theory of truth; by giving an account of truthmaking; by providing a suitable theory of reference; and perhaps in other ways. However it is done, a realist needs to make sense of the idea that theories yield knowledge of both observable and unobservable aspects of the world, and issues of correspondence unavoidably arise in this context. Realists cannot avoid talking about correspondence in some form, even if it is by means other than the correspondence theory of truth. Thus, Giere cannot dissolve the issue of correspondence simply by embracing the redundancy theory of truth instead.

In order to see more clearly why merely invoking the redundancy theory in connection with hypotheses about similarity relations between models and reality will not suffice for the realist, let us consider some of the different sorts of "correspondence" at issue when models are employed to represent parts of the world. Firstly, one might say that certain linguistic formulations correspond to specific models. This is just a loose way of noting that models are defined by their axioms, which no one would dispute. Secondly, one might say that models correspond to certain classes of phenomena. This too is uncontroversial, for it is simply a way of suggesting that scientific models are constructed with certain parts of the world in mind, *viz.* those target systems they are intended to represent. A third sort of correspondence concerns Giere's hypotheses. Whenever one claims that a similarity relation obtains between a model and the world, one might say that there is a correspondence between that claim and an actual similarity relation that obtains. In contrast with the two previous cases of correspondence-talk, there is a significant chance here that correspondence might fail. For there is correspondence in this third case if and only if the claimed similarity relation actually obtains (that is, if and only if the hypothesis is true), and this will not always be so. A fourth sort of correspondence may apply between descriptions and target systems. Correspondence obtains in this final case if and only if descriptions characterize aspects of reality well, meaning that they are true, that they refer, or what have you.

Here one comes to the heart of the matter. Constructive realists think that one can combine the unproblematic first case of correspondence between models and their axioms with the third case involving hypotheses or claims of similarity between models and the world, in such a way as to give an account of scientific knowledge. A redundancy theory of truth, they maintain, is all that is required where hypotheses are concerned. To say that it is true that a relation of similarity obtains is merely to say that a relation of similarity obtains. But this is insufficient for realism. A claim of similarity, even when given in terms of respects and degrees, does not by itself yield any information about the ontological details of which realists claim knowledge, and that theories are supposed to deliver. In order to understand clearly what a claim of similarity is telling one about the world, one must interpret the claim that some description is true, both of a model and of reality. It is the second conjunct here that most interests realists, and this is precisely where the fourth case of correspondence comes into play. Here the application of a description to the world must be interpreted, for descriptions, like all representations, are generally susceptible to different interpretations, and thus ambiguous otherwise.

Do the values for velocity generated by the two-particle Newtonian gravitational model merely correspond to numbers displayed on the output screens of scientific devices, or do they also correspond to the velocities of planets and satellites as conceived by realists? Logical empiricists, constructive empiricists, realists, and other epistemic agents interpret claims of similarity between scientific models and the world differently. The semantic view takes a detour from language via models, but in order to understand what hypotheses are telling one about the world one inevitably returns to issues of correspondence in connection with language. Access to the information models contain regarding the ontology of a particular or a process simply cannot be had unless one interprets descriptions of these non-linguistic representations. Realists make substantive claims as to whether the properties, relations, and particulars represented by models have counterparts in an external reality – whether, for example, the inverse square relations represented in Newtonian models actually obtain in the world. Claims of similarity between models and reality offer no escape from this practice, because their inherent ambiguity is dispelled only insofar as they can be interpreted. That is, they are helpful only insofar as they yield descriptions of aspects of the world that can be interpreted in the manner of realism, or empiricism, and so on.

Consider once more the constructive realist's example of models of Newtonian gravity. Contemporary physics holds that Newton's theory is not true, strictly speaking, though terrifically useful and predictively accurate to an impressive degree. If there are elements of truth contained within past (and present) theories, these are hopefully subsumed into theories that succeed them. Realists commonly contend, for example, that when earlier theories describe well-detected relations, later theories generally retain these descriptions or apply them to limiting case situations. These sorts of claims, however, cannot be entertained merely by asserting similarities between models and their target systems unless clarifying interpretations are added and considered. These considerations are indispensable to realism, for they are indispensable to the project of spelling out precisely what similarities obtain and how. The relation of similarity cannot bypass issues of correspondence, as Giere hopes, for mere assertions of similarity underdetermine ontology. Uninterpreted, these assertions are ambiguous and thus incapable of expressing substantive epistemic claims. Spelling them out produces descriptions of aspects of models that can be interpreted to yield knowledge of the world.

Another virtue envisioned by the constructive realist of using hypotheses to assert similarity relationships concerns the issue of reference for theoretical terms. Recall the pessimistic induction (PI) on past science, according to which it is likely that contemporary theories are false, by induction on past theories now regarded as false from the perspective of the present. One popular way of formulating PI focuses on the failure of reference of various theoretical terms that were central to past theories. 'Phlogiston', 'caloric', and 'the optical ether', for example, were thought to refer at one time, but all are now construed as non-referring. Constructive realists, however, suggest that their version of the semantic view furnishes an antidote to worries about reference. Since models need only be similar to parts of the world, the non-existence of various referents of past theory terms ceases to be a concern. As Giere (1988, p. 107) notes: 'Whether the ether exists or not, there are many respects in which electromagnetic radiation is like a disturbance in an ether.'

If the existence of specific properties, structures, and particulars were unimportant to realism, and if realists had no interest in distinguishing their views from those entailed by different epistemic commitments, then certainly PI would no longer constitute much of a worry. A knowledge of these aspects of the world, however, is precisely what realists are interested in, and realism must be distinguished, for example, from an epistemic commitment to observables only. A semirealist may interpret the evidence of well-detected relations as yielding information about unobservable causal properties and concrete structures. To assert these beliefs and to distinguish them from the beliefs of other epistemic agents who would endorse the very same claims of similarity between models and the world, the realist interprets descriptions. Giere goes on to say that one good reason for rejecting ether models is the fact that there is no ether, and that this constitutes an important respect in which similarity between these models and reality fails to obtain. But now constructive realists surely want to have their cake and eat it too. For if one spells out claims of similarity in such a way as to consider whether and to what extent aspects of models (representations of properties, relations, and things like the ether) have counterparts in reality, one engages in the very project of interpreting descriptions that constructive realism is supposed to disavow.

Similar difficulties will confront the constructive realist version of the model approach wherever realists are required to interpret descriptions. For example, those interested in matters of scientific methodology might wonder how realists should understand the nature of theory choice. On the semantic view this issue turns on the question of what criteria are used in order to determine which families of models best fit the world. But again, in order to avoid ambiguity one must consider what 'fit' means in this context. If one construes it in terms of making claims about similarity relations between aspects of models and the world, one inevitably confronts the dilemma I presented to the realist adopting the semantic view. Realists must either spell out hypotheses in terms of descriptions that can be interpreted, or give up their commitment to any realism worthy of the name. Good theories contain information about the things they represent, and in extracting this information one inevitably confronts issues of correspondence between language and the world. Non-linguistic representations play a crucial role in the sciences, but they cannot be used epistemically to sidestep challenges associated with linguistic interpretation.

7.5 APPROXIMATION AND GEOMETRICAL STRUCTURES

I have now examined two of the more fully developed accounts of the semantic approach to theories by authors who suggest that an emphasis on models facilitates a commitment to some form of realism. To conclude this discussion, let me consider one final proposal in the spirit of this suggestion. An investigation of this proposal will also provide some background for the topic of approximate truth, which is the unifying theme of Chapter 8. Peter Smith (1998) offers what can be described as a geometrical version of the semantic view, limiting his discussion to the specific context of dynamical theories. A dynamical theory is defined here as one that describes how the values of certain parameters evolve over time, but excludes representations of mechanisms that are relevant to the question of why such evolutions take place. That is, a dynamical theory ostensibly excludes details that might otherwise serve in causal explanations of the evolution of the process it describes. Smith thinks of dynamical models as abstract objects defined by specific geometrical structures. The degree of empirical success a theory has can be explained on this view by the extent to which the relevant geometrical structures approximate ones that can be associated with the dynamical system in the world it represents. The basic idea is that relations between various parameters in both target systems and models (concrete structures and their representations, to the semirealist) can analysed as having geometrical structures, which can then be compared.

Perhaps the best way to illustrate the idea of a geometrical semantic view is with an example, and Smith (1998, pp. 259–60) helpfully provides one:

> Consider the familiar account of the dynamics of a freely swinging pendulum. One standard way of looking at this account is to regard it as first characterizing a pure abstraction, the ideal frictionless pendulum moving in a plane according to Newton's laws. The governing equations determine the allowable patterns for the time-evolution of the ideal pendulum's angular displacement and velocity as a function of the pendulum's fixed length, etc. If we conceive of plotting a three-dimensional graph of time against displacement against velocity, then a certain bundle of three-dimensional curves will trace the allowable behaviours of a pendulum of given length subject to a given force. If we conceive, yet more abstractly, of these three-dimensional bundles being 'plotted' against pendulum length and applied force, we will get a more complex five-dimensional structure that in addition encodes the way that the possible behaviours of the pendulum depend on the length and force.

It is important to note here that Smith's notion of a geometrical structure differs from the notion of a concrete structure I described earlier in the exposition of semirealism. I analysed the latter concept in terms of certain relations between specific kinds of causal properties (and thus, consequently, between various particulars). A geometrical structure, on the other hand, is something rather different. It is a property of a representation of a concrete structure. Smith's description of the dynamics of the pendulum provides a striking example of how properties and relations composing concrete structures can be represented graphically. These representations or graphs have characteristic geometrical shapes, and this is what Smith intends by the term 'geometrical structure'.

One of the primary goals of Smith's discussion is to develop a model-theoretic account of approximate truth applicable to dynamical theories. The task of making sense of the idea of approximate truth more generally is a long-standing challenge to realism, and Smith (1998, p. 275, n. 30) is careful to skirt the issue of whether his account satisfies this particular challenge. Nevertheless, he does suggest that his proposal should be congenial to both realism and antirealism. Since it is intended to be acceptable to realists, and since many hold that having a concept of approximate truth would be an important fillip to realism, let us consider Smith's proposal in the same manner as the previous examinations of quasi-realism and constructive realism. That is, let us consider the question of whether Smith's account conceived as a geometrical version of the semantic view offers any special facilitation of realism. As we shall see, in some cases (though not in others) this proposal may furnish an attractive means of assessing the degree to which representations accurately describe their subject matter. As in the cases of quasi-realism and constructive realism, however, this potential benefit cannot be enjoyed in a purely non-linguistic manner. Realism requires that a model-based geometrical account of approximate truth incorporate interpretations of linguistic entities, and this will return us to the now familiar theme of contravening the spirit of the semantic view.

Realists generally accept that most theories past and present are false, strictly speaking, but they also generally contend that theories within specific domains get better over time, and not merely in terms of the accuracy of their observable predictions. According to realists, theories get better in terms of the accuracy of their descriptions of the natures and behaviours of things in the world, observable and unobservable. Hence the motivation for an account of approximate truth with which to "measure" improvements in the information theories contain, or to give a relative ordering of theories within a domain with respect to their descriptive closeness to the truth. Smith's proposal is addressed to precisely this sort of consideration. A wide class of dynamical theories, he says, can be thought of as containing two parts: one specifying an abstract geometrical structure as explicated a moment ago, and the other giving empirical application to that structure via the claim that it approximates a geometrical structure associated with some dynamical system in nature. Approximating truth for dynamical theories is thus a simple matter of approximating geometrical structures, and the closeness of such approximation can be measured mathematically quite easily.

What does it mean, however, to say that the geometrical structure of a model approximates that of a system in the world? As the example of the pendulum indicates very clearly, it would be a confusion to think that one can compare a geometrical structure instantiated by a model with one instantiated by something in nature. For recall, geometrical structures are properties of representations. In the example of the pendulum, these structures are properties of functions of bundles of curves in a graph, and such things are not present in the world, but aspects of representations. Nevertheless, one may charitably understand the requisite notion of approximation here in terms of comparing a geometrical structure instantiated by a model with one that is instantiated by another model – a model of the data, constructed using observed or detected values of the relevant parameters. In this case, if the geometrical structures are close, this entails that the values of the parameters whose functions are graphically represented to produce these structures are also close. And this entails that one's model yields accurate descriptions of the data, within some specified measure of accuracy. But note, once again: this by itself makes no commitment with respect to ontology. Realists maintain that theories often furnish descriptions of unobservable properties and relations, but mere geometrical closeness, much like a hypothesis on Giere's view, underdetermines these ontological details. Geometrical closeness is compatible with very different epistemic commitments including realism and empiricism.

Of course, the fact that geometrical closeness is open to interpretation by different epistemic agents is no problem for a geometrical version of the semantic view. Indeed, Smith acknowledges this feature of his position when he claims that this approach to approximate truth should be congenial to both realists and antirealists. It is a problem, however, for anyone who might hope to use Smith's proposal to suggest that focusing on models can help realists to bypass the interpretation of descriptions. Though the geometrical features of models may prove useful in assessing the relative approximate truth of some theories, this does not suggest that a model-theoretic approach dissolves the need to consider issues of correspondence. The moral here once again is that realism cannot avoid linguistic entities by focusing on non-linguistic representations. A realist and an instrumentalist, for example, might agree that a dynamical theory is approximately true in Smith's sense, but in most cases they will have very different conceptions of what this means. Since rival interpretations are consistent with one and the same instance of geometrical closeness, and since the approximate truth of a theory here can mean very different things to different epistemic agents, the only means by which to disambiguate a claim that a model is approximately true, or more approximately true than another, is to interpret descriptions of the properties and relations it represents.

Many interpretations of the information a model yields regarding the world are possible, and like anyone else a realist on the semantic view faces the challenge of specifying one of them. This challenge is greatly reduced in connection with at least one class of theories, however. If the properties and particulars represented by a family of models are all observable, a realist interpretation of the theory and the interpretation of many empiricists will coincide, since there are no putatively unobservable properties and particulars about which to disagree. In this case it is arguable perhaps that geometrical closeness by itself, in the absence of any interpretations of linguistic devices, can serve as an indicator of approximate truth for anyone who makes an epistemic commitment to the observable parts of the world. In such a case and among these epistemic agents, geometrical closeness might well indicate approximate truth *simpliciter*. In the sciences, however, relatively few theories fall into this class, and the most interesting and controversial cases for realists and empiricists are theories that do not. Any model that incorporates elements whose putative counterparts in reality can be interpreted as unobservable properties and particulars will be understood differently by people with different epistemic commitments regarding them. In order to specify any one such understanding, descriptions of models must be interpreted so as to clarify what sort of knowledge of the world is intended.

For the semirealist, talk of geometrical structures is at best a kind of shorthand for finer-grained claims regarding properties, whose magnitudes and relations are represented using models to create geometrical structures in the first place. As I have said at various points earlier, semirealism is a realism about causal properties and their relations in the first instance. And while a consideration of geometrical structures may prove a useful tool in some cases, it is questionable whether geometrical closeness is co-extensive with what sophisticated realists would recognize as approximate truth more generally. Earlier I suggested, for example, that the issue of whether aspects of models are abstractions or idealizations is relevant to a realist appraisal of theories, but approximate truth on Smith's account is not sensitive to this distinction. For instance, in some cases a model may offer a high degree of geometrical closeness with respect to certain parameters despite incorporating a high degree of idealization. If there are such cases, as I will suggest momentarily, then geometrical closeness is not a sufficient condition for approximate truth in the realist sense. Perhaps it is best understood as one criterion that can be weighed in combination with others, such as the extent to which a theory abstracts or idealizes, in determining whether a realist should recognize it as approximately true.

Let us consider one last example, also discussed by Smith (1998, pp. 274–5), which illustrates the possibility of a divergence between his concept of approximate truth and one that would serve realism. Variations on the Ptolemaic system of astronomy (which Ptolemy originally proposed in the second century CE) were not wholly supplanted until some time after Copernicus' dramatic rejection of them in the sixteenth century. Smith considers a thought experiment in which a version of the Ptolemaic theory is compared to a Newtonian theory. Imagine that T_1 and T_2 are Ptolemaic and Newtonian dynamical theories of planetary motion, respectively. It turns out that the parameters of T_1 and T_2 can be chosen in such a way that T_1 is more approximately true than T_2 in Smith's sense. That is, T_1 is superior in approximating the geometries mapped out by planetary motion; consequently, T_1 is better at predicting the motions of planets than T_2. There is something uncomfortable about this, however. Most realists regard Newtonian theory as improving on, and thus more approximately true than, Ptolemaic theories.

Smith suggests two ways out of this conundrum, but neither, I believe, is satisfactory for realism. The first is to think of judgments of approximate truth as interest relative, or dependent on the specific uses to which a theory is put. If one's interests are purely navigational, for example, one might accept that T_1 is more approximately true in this respect. There is a sense in which this is undeniable, for all that is being asserted is that T_1 yields more accurate predictions of planetary motion than T_2. As an account of the approximate truth of theories, however, this will not serve the realist. Tying the notion of approximate truth too closely to the predictive uses to which theories are put is fitting only for versions of instrumentalism and empiricism. Instrumentalists and realists both will of course regard T_1 as a better tool than T_2 for the task of navigation, but unlike an instrumentalist, the realist must take more (such as idealization) into consideration than merely observable predictions in thinking about approximate truth. Indeed, not only does geometrical closeness not entail approximate truth for the realist, but a lack of geometrical closeness does not entail that a theory is false. According to semirealism, abstract theories may yield true descriptions of properties and relations even if their predictions are rather inaccurate in some contexts. Abstract theories may furnish excellent descriptions of circumstances in which only the concrete structures they describe are present, but these same descriptions may fare poorly in other circumstances.[3]

Smith goes on, however, to suggest a second strategy for reconciling realism with the fact that a Ptolemaic theory might score better than a Newtonian theory with respect to geometrical closeness. What if T_2 could be unified with a greater number of more approximately true theories than T_1? In this case, one might suggest, despite the fact that T_2 does not approximate geometrical structures associated with planetary motion as well as T_1 does, there is a basis for the claim that T_2 is more approximately true. Many realists do think that in some cases, theory unification is an

---

[3] These considerations stand in contrast to views according to which approximate truth can be assessed only relative to specific contexts of use; cf. Teller 2001b, pp. 402–4. In Chapter 8, I will outline an account of approximate truth that is not hostage to interests or purposes.

indicator of truth, but I will not consider the merits of this idea here. For even if there is virtue in unification, however it is understood, this strategy cannot assist in the present task of furnishing a satisfactory account of approximate truth. If T_2 can be more approximately true than T_1 despite the fact that it fares less well on Smith's criterion of geometrical closeness, then clearly geometrical closeness by itself cannot be used to determine which other theories – those with which T_1 and T_2 can be unified – are more or less approximately true. In order to determine the relative approximate truth of these other theories, one would have to take into account the approximate truth of theories with which *they* can be unified, and this courts worries of circularity and regress.

To make judgments about the truth or relative approximate truth of theories, realists must interpret descriptions of aspects of models in connection with the things they represent, and this exceeds the use some model theorists might hope to make of Smith's proposal. This brings us back once more to the issues that prompted my consideration of the implications of the semantic view for realism. In order to differentiate their epistemic position from other possibilities, realists must generally do more than merely hold models up against the world. Non-linguistic representations may contain information about aspects of their target systems, but there are many ways of interpreting them. Epistemic commitments are specified using models only when descriptions of them are formulated and interpreted, and this opens the door to the very issues of correspondence some realists hope to avoid by emphasizing models. The semantic view seeks to separate theories from language, but no epistemic commitment can be entertained on too strict a separation. To think otherwise does not take seriously the question of how these commitments are made. And so, even if the semantic approach is a perfectly acceptable view of the nature of theories, it does not make realism any easier than it might otherwise be, and any suggestion to that effect is likely deceptive. So far as realism is concerned, the issue of what a theory *is*, "ontologically" speaking, is likely a matter of convention. Models have important uses, but the metaphysical nature of scientific theories has no implications for the epistemology of the sciences, and *vice versa*.

CHAPTER 8

Approximate truths about approximate truth

8.1 KNOWLEDGE IN THE ABSENCE
OF TRUTH *SIMPLICITER*

I began this book with the question of what scientific realism is, suggesting in the first chapter a very rough first draft of an answer: it is the view that our best scientific theories give approximately true descriptions of both observable and unobservable aspects of a mind-independent world. In the many pages since then, I have attempted to refine this answer in several ways. In response to the further question of what more precisely a sophisticated realist should believe in, I laid out an account of causal properties, concrete structures, and particulars described by theories under epistemically favourable conditions. In response to the challenge of furnishing a coherently integrated understanding of the key metaphysical concepts commonly associated with realism, I developed a unified picture of causal processes, laws, and kinds, in support of semirealism. And in response to the question of how exactly the realist should cope with the facts that human beings are fallible, and that epistemic conditions are often far from favourable, I have suggested throughout that plausible forms of realism are generally selective with respect to the parts of theories they endorse for belief, and that their commitment to even these parts is inevitably graded, reflecting a range of degrees of causal contact with the properties and structures they putatively concern.

But how are realists to reconcile the idea of a selective endorsement of parts of theories with the notion of theoretical truth? Given the optimistic yet partial nature of their belief in most cases, it seems clear that by their own lights, semirealists are neither interested nor entitled to describe most scientific theories as true *simpliciter*. The term 'approximate truth' shoulders a heavy burden here, and antirealists have not been shy in pointing this out. In Chapter 7, I hinted at several issues that need to be considered in connection with the idea of truth in scientific contexts, and it is time now to explore these issues in more detail. The present challenge to realism is to give an account of knowledge that is capable of embracing theories that are often and perhaps even typically false, strictly speaking. In this final chapter, I will suggest that work on this question to date, though technically ingenious, does not adequately explore the qualitative details of how sophisticated realists should understand the ways in which scientific theories deviate from the truth. My primary goal here is to outline those details, and to consider how they should inform realist assessments of the epistemic status of scientific theories. The core of this outline is the idea that approximate truth for the realist is something properly conceived in heterogeneous terms: it is multiply realized by means of different sorts of representational relationships between theories and models on the one hand, and things in the world on the other. These different forms of representation reflect the degrees to which theories and models abstract and idealize, and as a consequence, I will suggest, a realist understanding of approximate truth should be sensitive to these different manners of representation.

To begin, let me recall a summary of the familiar motivation for an account of approximate truth shared by all versions of scientific realism. Realists accept that most theories past and present are false, strictly speaking, but nevertheless contend that subject to the various caveats I have endeavoured to make transparent, they yield knowledge, and that within specific domains of scientific investigation theories get better over time, not merely in terms of the accuracy of their observable predictions. Theories get better, realists contend, in terms of the accuracy of their descriptions of the natures and behaviours of target systems in the world, observable and unobservable. Hence the motivation for an account of approximate truth with which to "measure" improvements in the infor- mation theories contain, or to give a relative ordering of theories within a domain with respect to their closeness to the truth. In tackling these issues, however, it should be understood that success cannot be measured in terms of a resultant ability to deliver precise assessments of approximate truth on demand. In order to characterize very precisely the manner in which a given scientific theory deviates from the truth, one must know the true theory from which the given theory deviates, and if one knew all true theories, there would be little controversy over the notion of approximate truth! The goal of these considerations is to respond to antirealist scepti- cism by demonstrating that approximate truth is, in fact, a coherent idea.

I will collect previous work on this topic into three groups, corre- sponding to the main extant approaches to the notion of approximate truth. In none of these cases will I be concerned to furnish an exhaustive evaluation of their merits and defects; the goal is rather to convey a basic sense of the strategies they adopt, thereby setting the stage for an outline of some further and epistemically important considerations to follow. To introduce some convenient labels, I will refer to these previous accounts respectively as the verisimilitude approach, due to Karl Popper; the possible worlds approach, elaborated in different ways by several authors including Pavel Tichý, Ilkka Niiniluoto, and Graham Oddie; and finally the type hierarchy approach, due to Jerrold Aronson, Rom Harré, and Eileen Cornell Way. Let us consider these views in turn.

8.2 MEASURING "TRUTH-LIKENESS"

Popper was the first to propose a clear definition of what he called 'verisimilitude', or "truth-likeness". On his (1972, pp. 231–6) view, scientific theories within a domain may exhibit increasing levels of verisimilitude over time, and this relative ordering can be expressed in a simple, intuitive way. Imagine a sequence of theories occurring within a particular scientific subdiscipline, all putatively concerning the same subject matter – T_1, T_2, T_3 . . . – ordered temporally over the history of that field. Now, for each theory in the sequence, consider the set of all of its true consequences (for example, T_1^T), and the set of all of its false consequences (T_1^F). A comparative ranking of the verisimilitude of any two theories in the sequence can be given, suggested Popper, by simply comparing how much truth and falsity they each contain, by means of comparing their true and false consequences. For any given theory, T_n, and any previous theory in the sequence, $T_{<n}$, T_n has a higher degree of verisimilitude than $T_{<n}$ if and only if either of the following statements is true:

1. $T_{<n}^T \subset T_n^T$ and $T_n^F \subseteq T_{<n}^F$
2. $T_{<n}^T \subseteq T_n^T$ and $T_n^F \subset T_{<n}^F$

The symbol '$\subseteq$' stands for set-theoretic inclusion (X is a subset of Y if and only if Y includes X), and the symbol '$\subset$' stands for proper inclusion (X is a proper subset of Y if and only if Y includes X and X and Y are not identical). To put it less elegantly but in words, a later theory has a higher degree of verisimilitude than an earlier one if and only if one of the following two conditions is met: the later theory has all of the same true consequences as the earlier one plus more, and the same or only some of the false consequences of the earlier theory; the earlier theory has the same or only some of the later one's true consequences, and all of the same false consequences as the later theory plus more.

The idea of comparing the contents of different theories this way so as to ascertain their relative approximate truth is, no doubt, highly intuitive, but sadly Popper's account is subject to well-established and fatal difficulties, first described by David Miller (1974) and Tichý (1974). These authors proved independently that in the case of neither 1 nor 2 above can both conjuncts be satisfied together in a comparison of false theories. It turns out that on Popper's definition of verisimilitude, in order that T_n have greater approximate truth than $T_{<n}$, T_n would have to be wholly and completely true. Thus, on the verisimilitude approach, one false theory can never have more approximate truth than another, and this rather defeats one of the purposes I have assumed here on behalf of scientific realism, to furnish a coherent account of what it means for false theories within a domain to be increasingly approximately true. It is generally held that neo-Popperian attempts to define verisimilitude in similar terms meet similar fates.[1]

A second family of accounts concerning the notion of approximate truth is what I labelled the possible worlds approach (also sometimes called the 'similarity' approach). The precise differences between different versions here are interesting but inconsequential for my purposes, so let me give a rather general characterization. The basic strategy of the approach is first to identify the truth conditions of a theory with the set of possible worlds in which it is true. Then, one calculates what many of the proponents of this view call truth-likeness, in terms of a function that measures the average "distance" between the actual world and the worlds in that set, thereby generating an ordering of theories with respect to truth-likeness in terms of these distances. One way to do this is to consider the class of atomic propositions entailed by a theory, each attributing a specific state to a particular. Possible worlds are then described by distributions of truth values across these atomic propositions. The greater the extent to which a given theory agrees with a theory correctly describing the actual world, the greater the given theory's truth-likeness.[2]

Let me mention *en passant* two of the more important controversies surrounding the possible worlds approach. Perhaps the best-known

---

[1] For example, see Oddie 1986a for a discussion of Newton-Smith's neo-Popperian account.

[2] See Tichý 1974, 1976, 1978, Niiniluoto 1984, 1987, 1999, and Oddie 1986a, 1986b, 1990. Niiniluoto 1998 provides a helpful and comprehensive summary of different authors' contributions to this approach.

objection is due to Miller (1976), who argues that on this view, both measures and relative orderings of truth-likeness are language-dependent, which is something no properly "objective" account should allow. That is, logically equivalent theories may have different degrees of truth-likeness depending on the language in which they are expressed, and the relative truth-likeness of two different theories may be reversed when translated into another language employing logically equivalent predicates. Aronson (1990) points out that on the possible worlds approach, the truth-likeness of a proposition, whether true or false, depends on the number of atomic states under consideration. The truth-likeness of true propositions decreases, and the truth-likeness of false propositions increases, as the total number of states described by a theory increases. It is certainly questionable, however, whether the truth-likeness of a proposition should vary as a function of the total number of propositions concerning states of affairs *other* than that described by the proposition at issue. Aronson (1990, p. 9) describes this as 'a pernicious holism, one where the verisimilitude of a proposition becomes overly dependent on the truth of other, completely irrelevant, propositions (or states)'.

I will not consider these charges in any detail here, but let me at least note that proponents of the possible worlds approach dispute their force. Tichý (1978), for example, agrees that measurements of truth-likeness are relative to what he calls 'epistemic frameworks', defined by the sorts of objects and properties one associates with a language, but argues that one cannot, *contra* Miller, translate between such frameworks.[3] Niiniluoto (1984, p. 166) also accepts that truth-likeness is relative to what he calls 'cognitive problems'. Imagine a set of 'alternative states of nature', S, each element of which represents a possible state of something in the world; only one element of S, s^*, obtains in the actual world. A cognitive problem takes the form of a question concerning which state in S is the actual state s^*. Crucially, he contends, judgments of truth-likeness and relative truth-likeness should be restricted to cognitive problems expressed in languages that employ natural kind predicates, as opposed to the "gerrymandered" predicates of supposedly inter-translatable languages, such as those suggested by Miller. As Oddie (1986b, p. 159) puts it: 'Any rebuttal of the Miller argument, whatever its semantic presuppositions, must grant certain properties, magnitudes, or constants, a privileged status.' In response to the

---

[3] Tichý 1976, p. 35, claims that truth-likeness is relative to a 'logical space', understood as the totality of functions taking atomic propositions to truth values. Tichý 1978 redefines the idea of a logical space as the totality of possible worlds relative to an epistemic framework.

charge of pernicious holism, one might note that Niiniluoto's (1999, pp. 73–4) account incorporates an "informational" component, suggesting that measures of truth-likeness *should* vary depending on the amount of information an atomic proposition contains relative to the total number of states under consideration.

I have nothing to add presently regarding the debate concerning holism and information, but leaving aside the issue of natural kinds for the moment, there are grounds for worry here about the response of the possible worlds approach to Miller's argument concerning language dependence. The development of new scientific theories often involves the introduction of new concepts (and the rejection of previous ones) – concepts that require a language different from and sometimes richer than that which sufficed for the expression of its predecessor. As Niiniluoto (1984, p. 167) admits: 'If the formulation of the problem turns out to be insufficient or inadequate, we have to proceed to consider *deeper* cognitive problems by extending the language L [the language used to describe the states in S] with new vocabulary.' But in order to compare the truth-likeness of two theories, they must have the same epistemic framework or address the same cognitive problem, since as I noted a moment ago, on the possible worlds approach, measures of truth-likeness are relative to frameworks or problems. Even assuming that a theory can be translated from its own language (L) into the language of a subsequent theory (say, some extension of L), there is no reason to expect that later theories will generally preserve the epistemic frameworks or cognitive problems of their predecessors, as Niiniluoto concedes.

The third and final account of approximate truth I undertook to survey is the type hierarchy approach. Aronson 1990 and Aronson, Harré, and Way 1994, pp. 15–49, characterize verisimilitude in terms of similarity relationships between nodes in type hierarchies: tree-structured graphs of types and subtypes. The nodes in these graphs represent concepts or things in the world, and links between nodes represent relations between concepts or things. Consider, for example, a taxonomy of organisms divided into kingdoms, each of which is further divided into phyla, and so on into classes, orders, families, genera, and species. Similarity is then defined with respect to locations within type hierarchies. Given a hierarchy of aquatic animals, for instance, in order to determine that dolphins are more similar to whales than to tuna, one calculates their degrees of similarity to one another in terms of a weighted difference measure generated by a comparison of the properties these types have in common and those in which they differ. Now, imagine the same sort of comparison, this time between a node in a theoretical type hierarchy and a corresponding node in the actual type hierarchy of the world (or more pre-cisely, that part of the world the theory describes). Verisimilitude is correlated with the "distance" between a theoretical claim about a type and the correct description of that type, reflecting the degree of similarity of the nodes with which they are associated.

Perhaps the most serious difficulty with this proposal is that, just like the possible worlds approach before it, the type hierarchy approach makes an important but dubious assumption regarding natural kinds. Aronson's most common examples concern the verisimilitude of claims about type or natural kind membership, such as 'x is a dolphin', where x stands for a particular belonging to some specific natural kinds. Unless there is only one objectively correct type hierarchy of the contents of the natural world, however – a position strongly rejected in Chapter 6 – there can be no determinate answer to the question of what a node in a theoretical type hierarchy should be compared *to*. Different and entirely legitimate scien-tific investigations and consequent judgments regarding which similarity relationships between things in reality should be privileged may lead to different, entirely valid type hierarchies (cf. Psillos 1999, p. 277). Robbed of the notion of "the correct" type hierarchy of the world, the concept of similarity suggested by the type hierarchy approach is undefined, and as a consequence, so is its concept of verisimilitude.

The preceding survey of accounts of approximate truth has been quick and partial, and my goal in providing it has not been to furnish an exhaustive discussion, but rather to set the stage for some reflections to follow. I believe the difficulties reviewed and suggested here for extant accounts are serious, and in some cases fatal, but in the next section I will leave these disputes to one side. In order to understand better the nature of approximate truth and its connection to scientific realism, I think it is important to have a deeper understanding of the qualitative dimensions of the concept than any of these previous accounts aspires to provide, concerning the ways in which theories and models typically diverge from truth in the first place. And in an effort to move towards such an understanding, I will take some inspiration from a surprising and perhaps unlikely source, *prima facie*: analogies between practices of representation in the sciences, and in art.

8.3 TRUTH AS A COMPARATOR FOR ART AND SCIENCE

Most work on the subject of approximate truth is formal in nature. That is, it attempts to analyse the concept in terms of precisely specifiable algorithms that yield mathematical measures of degrees of truth-likeness.

The difficulties facing these approaches are outstanding puzzles, but I believe that most realists operate with the view that an adequately articulated *informal* account of the concept of approximate truth would serve the project of establishing the overall coherence of realism perfectly well. This may be an implicit and widely held view, but it is fair to say that detailed articulations are thin on the ground.

Most realists, I think, would approve of Psillos's (1999, p. 277) informal characterization so far as it goes: 'A description D . . . is approximately true of [a state] S if there is another state S' such that S and S' are linked by specific conditions of approximation, and D . . . is true of S'.' But this is mostly a statement of the relevant intuition; it is not yet much of an explication of the concept of approximate truth. In the absence of further detail, the explication this statement provides yields little insight, for the explicans invokes the notion of 'conditions of approximation', which seems no clearer at first glance than the notion it is intended to explicate. In order to advance their understanding of approximate truth, realists require some further information regarding the nature of these conditions. In the remainder of this chapter, I will attempt to explicate the concept of approximate truth informally by shedding some light on the notions of approximation at issue. I will suggest that there is no reason to expect any one form of approximation to be relevant to all cases of scientific theorizing, and that several related considerations help to illuminate the ways in which different forms of approximation are exemplified by theories and models.

The inspiration for the account to follow stems from work on the subject of representation in the philosophy of science, which has recently begun to draw on the resources of an important literature on representation in the philosophy of art. Much of this work is concerned narrowly with visual representation, whereas the goal of this chapter is to consider, from a realist perspective, the epistemic status of scientific representation in all of its guises – visual, linguistic, mathematical, model-theoretic, and so on. In methodological spirit, however, my strategy here will be very much the same. The approach is to see whether reflections on the idea of truth and cognate notions in the context of art can help to illuminate the concept of approximate truth in the context of the sciences. Like many contributions to the subject of representation in art and science, I will take as my starting point some groundbreaking work by Nelson Goodman.

In *Languages of Art*, Goodman (1976) presents an extensive analysis of the 'symbol systems' different forms of art employ so as to express their content. It is not until the very end, however, that he raises the question that most concerns me, regarding analogies between art and the sciences concerning truth (p. 262):

Have I overlooked the sharpest contrast: that in science, unlike art, the ultimate test is truth? Do not the two domains differ most drastically in that truth means all for the one, nothing for the other? Despite rife doctrine, truth by itself matters very little in science.

Of course, Goodman does not think that truth has *no* importance when it comes to the sciences. Rather, he maintains that it must be understood in a carefully qualified manner if its relevance is to be correctly appreciated. Truth by itself matters little, for one thing, because one may generate trivial truths at will, for instance, simply by making empirical observations. But scientists are interested in important truths, viz. ones that have scope and specificity that are appropriate to the inquiries they undertake, and that raise and answer questions of significance. Furthermore, and most importantly for my purposes, Goodman acknowledges that laws are seldom true, but in the face of sceptical worries such as the underdetermination of theory by data, scientists use criteria such as simplicity 'as a means for arriving at the nearest approximation to truth that is compatible with our other interests' (1976, p. 263). Ultimately, he says, truth must be understood in terms of 'a matter of fit' between theories on the one hand and facts on the other. And interestingly, just this sort of "fitting" is characteristic of the relationship between art and the world.

Thus, the common idea that sciences concern truth, and art concerns something else entirely, is misleading. For ultimately, truth in both domains should be understood in terms of approximating reality by means of representations. This is all very suggestive and somewhat vague, but unfortunately for the reader, that is the end of the book! Goodman does not say anything more about these analogies of fit, or approximating truth, between art and science, or what terms such as 'scope' and 'specificity' might mean in evaluating the importance of claims about the world. Perhaps he thought that all the relevant clarifications are implicit in the pages that precede his finale, and indeed, I believe that clarifications regarding the scientific side of this equation are implicit in the previous chapters of the current work, so let me now attempt to make these latter clarifications more explicit. Here is a preview of what is to come: I will suggest that two central features of scientific knowledge illuminate the very thing that realists need to understand better – the idea of conditions of approximation. It is here that analogies to representational practices in art may indicate helpful ways of thinking. The first of these features is the distinction between abstraction and idealization in connection with scientific theories and models, and this will be the subject of the rest of this section and the next. A second key feature concerns the nature and pragmatics of scientific practice, which I will consider in the final section.

In Chapters 5 and 7, I examined two ways in which theories and models typically deviate from the truth: abstraction; and idealization. (I will generally talk of theories henceforth, but it should be understood that models have the same characteristics, and I intend the same conclusions in connection with them, *mutatis mutandis*.) Roughly, an abstract theory is one that results when only some of the potentially many relevant factors present in a target system are taken into account. Here one ignores other parameters, either intentionally for practical reasons, or unwittingly for reasons of ignorance, that are potentially relevant to the nature or behaviour of the system. An example I used earlier is the model of the simple pendulum, where, among other simplifying assumptions, one simply neglects frictional resistance due to air. In contrast, an idealized theory is one that results when one or more factors is simplified, again either intentionally or unwittingly, so as to represent a system in a way it could not be. Here one does not exclude parameters *per se*, but rather characterizes parameters that *are* taken into account in such a way that these characterizations are false descriptions of their intended referents, not least because properties and relations satisfying such descriptions are ruled out by laws governing the actual world. In the *Principia*, for instance, Newton assumes that the sun is at rest in his derivation of Kepler's laws of planetary motion. According to his theory, however, that would require that the sun be infinitely massive – not something that Newton believed. The sun experiences small amounts of motion due to the attractions of other bodies; the assumption that it is at rest is an idealization.

Niiniluoto (1999, pp. 136–8) mentions several examples of what I call abstraction and idealization. In correspondence, he suggests that both practices can be analysed mathematically in the same way, by letting the value of some parameter (such as air resistance) approach a specified limit (cf. Swoyer 1982, pp. 219–20). This view is commonly found in discussions of modelling, and there is significant evidence in its favour. Idealizations such as point masses and point particles, for example, can be described mathematically as limit cases in which the magnitude of the property of volume goes to zero. As we shall see, however, the importance of the distinction between abstraction and idealization is conceptual, not a function of how one treats them mathematically. As I noted earlier, these practices are by no means mutually exclusive, and representations are often both abstract and idealized. But given that pure abstractions (those incorporating no idealization) can be true and idealizations cannot, perhaps it should not be surprising that they bear somewhat differently on the issue of truth-likeness. More specifically, I believe that this distinction interestingly informs the conditions of approximation that are relevant to scientific theories, so let us turn to this matter now.

On an informal account, the question of how realists should think about approximate truth, I suggest, boils down to the following: how should they think about truth in light of these two practices of deviation? It would seem that there is a straightforward answer in the case of pure abstraction, and a less obvious one in the case of idealization. When I first introduced the concept of abstraction in Chapter 5, the context was a discussion of *ceteris paribus* laws; I argued there that in cases of pure abstraction, theories correctly describe naturally possible (nomically possible) target systems. In other words, they correctly describe properties and relations that could and sometimes do exist in the actual world. This stands in stark contrast to the view inspired by and sometimes associated with Cartwright (1983, 1989), that abstract theories are false, and that only theories incorporating very little abstraction approach the truth. Though I take this view to be mistaken, the error involved is a natural one, since pure abstractions are often *used* in something like the manner of idealizations. That is, one may apply purely abstract theories to systems in the world they do not correctly describe. Even the model of the simple pendulum, which is not a *pure* abstraction, serves to illustrate this point: insofar as it abstracts by neglecting resistance due to air, it does not correctly describe the behaviour of pendulums in circumstances other than vacuums, but it is regularly applied to atmospheric pendulums nonetheless. In these applications, the neglect of air resistance acts something *like* an idealization. But it is not an idealization strictly speaking, since vacuums are possible.

This suggests a simple, informal articulation of the notion of approximate truth *qua* abstraction. Consider all of the causal properties and relations relevant to the nature or behaviour of a particular system or class of target systems. Degrees of approximate truth are determined with respect to abstraction by the extent to which theories incorporate these properties and relations. The greater the number of causally relevant factors built into an abstract theory, the greater its approximate truth. This means of assessing relative approximate truth in these cases does justice to the intuition that higher degrees of abstraction correspond to lesser degrees of truth, but without committing the error of failing to note that abstractions may yet describe some things correctly. A pure abstraction furnishes perfectly correct descriptions of a certain class of target systems while being more or less approximately true in application to others. In such cases, the conditions of approximation relevant to assessing approximate truth can be understood simply in terms of how much information a theory provides, or its comprehensiveness, relative to a specific kind of system, or class of systems. Thus, pure abstractions may be true in application to some concrete phenomena, and more or less approximately true in application to others they do not comprehensively describe.

In cases of idealization, however, the realist faces a rather different challenge in making sense of the idea of conditions of approximation. For here, one does not have the luxury of descriptions of causal properties and relations that could be manifested in a least some circumstances, as in cases of pure abstraction. Idealizations are descriptions of properties and relations that do not and cannot exist as described in any circumstances. The modality of 'cannot' here may be understood in terms of the account of *de re* necessity and possibility outlined in Chapter 5. Consider an earlier example: models in classical mechanics generally treat the masses of bodies as though they are concentrated at extensionless points, but given the nature of mass properties and the dispositions with which they are associated, masses cannot be concentrated this way in any world such as ours, where particulars with these properties exist. Idealizations are strongly fictional in a manner that exceeds the deviation from truth characteristic of pure abstractions. In Chapter 7, I suggested that all successful scientific representations have intentional content (they are about things), and contain information about the things they represent. In cases of pure abstraction it is clear what form this information takes, but the situation is more complicated in cases of idealization, where realists must grapple with the question: what information about the actual world is contained in a fiction?

The fictional nature of idealizations may be displayed by recalling my earlier illustration of the possible worlds approach, in which one considers the class of atomic propositions entailed by a theory, each attributing a specific state to a particular (for example, 'the mass of a proton is 1.6726×10^{-27} kg'). In cases of pure abstraction, one is at liberty to say that the greater the extent to which a theory yields true descriptions of things in the actual world, the greater the theory's truth-likeness. But idealized theories do not generally give true descriptions of atomic states of affairs; idealized descriptions of properties do not truly describe *any* actual world properties. Recall the holism that applies to networks of properties discussed in Chapter 5: if the identity of a causal property is determined by certain dispositions for relations with other properties, then the natures of causal properties taken as a whole are constituted by a network of potential relations. The natures of individual properties are thus linked to one another via loops of potential relations, such that if any one of the causal properties described by an idealization is not a member of the network found in the actual world, none of them is.

This may well generate another difficulty for the type hierarchy approach to approximate truth, where one calculates degrees of similarity between theoretical propositions and true ones by performing weighted difference measures in terms of the properties these propositions describe in common and those in which they differ. Idealized characterizations do not describe *any* causal properties in common with true theories, strictly speaking, because they correctly describe fictional properties, not actual ones. On the type hierarchy approach, it is arguable that some true descriptions of the sorts of properties one attributes to particulars in order to characterize their locations in type hierarchies must be contained within a given theory for the very concept of approximate truth to get off the ground, but idealized theories generally do not give true descriptions of this sort.

The conditions of approximation relevant to assessing approximate truth *qua* idealization must be understood differently than the conditions of approximation *qua* abstraction. So how is a realist to make sense of approximate truth in connection with idealization?

8.4 DEPICTION VERSUS DENOTATION; DESCRIPTION VERSUS REFERENCE

At this juncture let me return to Goodman for a bit of inspiration, and draw a first analogy to representation in art. A moment ago I cited the final pages of *Languages of Art* as suggesting that in both art and the sciences, successful representation is a matter of fitting or approximating things in the world. Let me now consider this suggestion further, beginning with an examination of what Goodman has to say about realistic and non-realistic representation. It is precisely this distinction, I will argue, that scientific realists must appreciate in order to understand the truth content of idealized theories and models. The nature of this content is rather different than in cases of pure abstraction, and considering it will provide a crucial insight into how different contexts of representation call for a flexible approach on the part of realists, in their attempt to explicate the concept of approximate truth.

In the early stages of the book, Goodman (1976, p. 34) raises an important question – 'what constitutes realism of representation' – and immediately furnishes what must appear a provocative, negative answer: 'Surely not ... any sort of resemblance to reality.' If one interprets 'resemblance' narrowly here to mean 'similarity in appearance', this might seem a strange claim regarding much art though not perhaps regarding science, where one hardly expects sets of equations to bear similarities in appearance to chemical compounds or populations of organisms, for instance. On a broader reading of 'resemblance', such as 'having some feature or features in common', the potential strangeness extends to the scientific case, where realists believe that aspects of our best theories and models do, in fact, share features in common with their subject matter, such as commonalities in structure. This puzzle of interpretation is resolved with the further information that for Goodman, realism of representation is by no means inconsistent with resemblance, in either of the senses just suggested; his point is rather that realism is achieved only in special circumstances, viz. those in which agents considering a representation are aware of, or acculturated with, the system of representation used to encode information about whatever it is that is represented.

Consider a realistic picture, painted in ordinary perspective and normal colour, and a second picture just like the first except that the perspective is reversed and each colour is replaced by its complementary. The second picture, appropriately interpreted, yields exactly the same information as the first. And any number of other drastic but information preserving transformations are possible. Obviously, realistic and unrealistic pictures may be equally informative; informational yield is no test of realism ... The two pictures just described are equally correct, equally faithful to what they represent, provide the same and hence equally true information; yet they are not equally realistic or literal ... Just here, I think, lies the touchstone of realism: not in quantity of information but in how easily it issues. And this depends upon how stereotyped the mode of representation is, upon how commonplace the labels and their uses have become. (Goodman 1976, pp. 35–6)

Goodman is a conventionalist about systems of representation; anything can represent anything, subject to appropriately specified conventions. And one and the same representation can be realistic or not, depending on whether the relevant conventions have been internalized by the viewer or user.

There are several tantalizing issues concerning the connections between conventionalism and representation that deserve attention here, but for present purposes the key point I would like to extract is simply the idea that different conventions of reading information from representations are central to how one understands that information. With this idea in mind, here is a preview of one of the principal conclusions of this chapter: if realists are to have a genuinely informative, informal account of what it means to say that scientific theories are approximately true, the distinction between abstraction and idealization should be understood in terms of different conventions of representation. This is because, as I have suggested already, the conditions of approximation that are relevant to assessing approximate truth differ, depending on whether one is considering a theory *qua* abstraction, or *qua* idealization.

One last point borrowed from Goodman's analysis of artistic representation will prove helpful in arriving at this conclusion. Goodman (1976, p. 5) holds that 'the core of representation' is denotation. That is, in order for x to represent y, x must be a symbol for, or stand for, or refer to, y. Symbols here include 'letters, words, texts, pictures, diagrams, maps, models, and more' (1976, p. xi). Denotation is simply a species of reference, and more specifically, one that points from representations to things represented. Here then, finally, is a first analogy of artistic representation that I believe scientific realists can take inspiration from: just as in the case of art, where successful representation can be a function of denotation, in the sciences, successful representation can be a function of reference, even when theories contain only idealized descriptions of actual world properties. Let us consider this suggestion in some detail.

Emphasizing reference is of course far from novel in discussions of scientific realism. In Chapter 2, I discussed the position of entity realism (ER), which holds that under conditions in which one has significant evidence of an ability to manipulate or otherwise causally interact with entities, one has good reason to believe that such entities exist. Crucially, ER can be cast as a response to challenges realists face in light of the history of science, which teaches that theoretical descriptions are likely to change over time. Hacking (1983, ch. 6), for example, contends that one may continue to refer to the same causal entity despite changes in the theories that describe it, and this provides a stable point around which realists can organize their knowledge claims regarding unobservables. Despite the fact that theories are false and apt to change, there are conditions under which one has good reason to think that certain unobservable terms refer, and will continue to refer. The importance of reference relations has never really shaped thinking about approximate truth, but they would now appear to be relevant to understanding the differential truth content of pure abstractions and idealizations. Insofar as *true* claims about causal properties can be extracted from idealizations, these are generally claims of successful reference, not the more detailed descriptions one may associate with cases of pure abstraction.

This is a delicate matter, however. The assiduous reader will recall that in Part I, I also argued that ER is problematic as it stands. I made the case there that this form of realism does not draw the line between things realists should and should not believe in quite the right place, because when one considers scientific knowledge, existential claims about entities and further claims about their relations are not easily separated. I also maintained that ER is too crude, because there is something anachronistic about suggesting that scientists from different periods in the history of scientific investigation into a specific entity all believe in the same thing. The realist's story must be more refined, told at the level of specific properties and relations on which existential claims are based, and that are likely to survive (if only as limiting cases) in theories over time. Despite these terminal problems, however, I made a point of extricating what I take to be the important lesson of ER for the realist, that degrees of belief in unobservable entities are generally and rightly correlated with the extent of one's causal contact with those entities. An impressive ability to exploit systematically the dispositions associated with a property gives the realist strong grounds for belief, and less impressive abilities rightly ground more attenuated belief. On the impressive side of this continuum, claims of reference are concomitantly strong.

We are now in a position to see what it means for one theory to be more or less approximately true than another *qua* idealization, and to contrast this with the case of pure abstraction. So far as truth is concerned, even the best idealizations contribute primarily existential claims. They are not all on a par, however, when it comes to the approximate truth of the more substantive descriptions they provide. Some idealizations approximate true descriptions of various properties better than others, and this is what realists should have in mind when considering their relative approximate truth. The idea of approximation here is usually specified mathematically, and we have encountered several examples of this already. By showing how the equations of Newtonian mechanics are limiting cases of relativistic equations, one defines mathematically how Newtonian descriptions of the relevant properties approximate those of special relativity. The ideal gas law assumes that molecules of gas are point particles and that there are no forces of attraction between them, but it is possible to take into account both the space occupied by molecules of gas and small forces of mutual attraction. Thus, while the van der Waals equation generates values for various properties that approach those given by the ideal gas law at lower pressures (larger volumes), it yields different, more accurate values at higher pressures (smaller volumes).[4] The van der Waals equation, over certain ranges of pressure, volume, and temperature, describes the natures of these properties and their relations more accurately than the ideal-gas law.

Recall the statement of realist intuition I credited to Psillos earlier, to the effect that a description is approximately true of a state if it can be 'linked by specific conditions of approximation' to a true description. At that stage it was unclear what 'conditions of approximation' might mean, and thus I undertook to shed some light on this idea, with the goal of generating a more satisfying informal explication of the concept of approximate truth. The pieces of the puzzle required are now in hand. When theories deviate from the truth regarding their target systems, they do so via abstraction, or idealization, or in many cases, both. As I have argued, insofar as theories are abstract, approximate truth may be gauged in terms of the numbers of relevant features of their target systems they describe, so that theories incorporating greater numbers of these features are more approximately true than those incorporating fewer. Pure abstractions give descriptions of properties and relations that are true *simpliciter* of certain classes of target systems, and yet may be more or less approximately true in application to others. The notion of approximate truth *qua* abstraction is thus simply the notion of comprehensiveness, and the relevant condition of approximation here is the extent to which the numbers of factors represented by a theory match up with those in the target systems to which it is applied.

The notion of approximate truth *qua* idealization is different, for here the concern is not the comprehensiveness of descriptions, but the accuracy with which they characterize the natures of the specific properties and relations they represent. Unlike pure abstractions, idealizations do not generally offer true characterizations of the properties they concern, but they do permit ontological claims, in virtue of successful reference. By reducing the number and magnitude of idealized assumptions – by de-idealizing – one describes target systems to a greater degree of approximate truth. Unlike the case of abstraction, however, where improving a theory is simply a matter of increasing the number of potentially relevant factors it represents, there is no reason here to expect that processes of de-idealization across the sciences should follow any common pattern. There are many ways of incorporating idealized assumptions into theories, and the ways in which one describes possible de-idealizations may vary in just the way idealizations do. But whatever these variations, idealized descriptions of

---

[4] See McMullin 1985 for a discussion of this case (p. 259) and others like it.

properties and relations, or what I earlier called concrete structures, may improve in ways that are determinable in specific instances. The notion of approximate truth *qua* idealization concerns the degree to which a description of a concrete structure resembles a true description, where degrees of resemblance are defined as appropriate in each case. The relevant condition of approximation here is not comprehensiveness, but degrees of descriptive accuracy regarding concrete structures represented.

Let me revisit the first analogy to representation in art one last time before finishing with a second. In viewing a painting or a sculpture, one may extract more or less information regarding the things it represents, depending on the extent of its realism. At one end of this spectrum is what Goodman labels realistic representation in art, or depiction. Here the viewer is sufficiently acculturated with some relevant system of representation to derive significant information about that which is represented. At the other end of the spectrum representations may convey very little information, but information nonetheless. Consider the representational content of paintings, for example. Pablo Picasso's *Guernica* (1937) is one of the most famous works of the twentieth century, not least because of its awesome representational force. Its subject is the bombing of the Basque town of Guernica by Hitler's and Mussolini's air forces, with the complicity of Franco, during the Spanish Civil War. Aspects of the work – figures of a bull, a dead baby in the arms of a screaming woman, a speared horse, the broken body of a soldier, etc. – represent various things with greater and lesser degrees of realism. The painting taken as a whole also has representational content; one thing it represents is the rising threat of European fascism (see Suárez 2003, p. 236). Insofar as it does this, however, it is not depictive, but merely denotative. It does not furnish much in the way of "description" beyond making an existential "claim" about the presence of a terrifying danger.

Scientific theories that pass the threshold of realist acceptance likewise yield information about their subject matter, but whether they do so by furnishing nothing but true descriptions of properties and their characteristic relations, or merely by approximating the concrete structures to which they successfully refer, will depend on how abstract and idealized they are. The contrast between depiction and mere denotation as a central feature of representation in art is an analogy for the contrast between true description and mere reference as a central feature of representation in the sciences. Greater approximate truth can be understood in terms of improved representations of the natures of target systems in the world, and this improvement can be spelled out along two dimensions: how many of the relevant properties and relations one describes (abstraction), and how accurately one describes them (idealization). This simple formula, I suggest, together with an understanding of the relevant conditions of approximation that accompany it, is precisely the explication realists require of the principal notions composing an informal approach to the concept of approximate truth.

8.5 PRODUCTS VERSUS PRODUCTION; THEORIES AND MODELS VERSUS PRACTICE

The time has come to close this chapter and book. The study of scientific realism undertaken here has traversed a broad range of topics, from the place of realism as a rival to empiricism in the philosophy of science, to an exposition of what I presented as the most promising face of realism today (semirealism), through a proposed, integrated account of the key concepts underpinning this realism including causation, properties, *de re* necessity, and natural kinds, and finally to epistemic considerations regarding the nature of scientific theories and models, and their truth content. The emphasis throughout has been on demonstrating that the many disparate elements commonly associated with a realist world view can be combined into a coherent, unified package. As I have stressed at several points, however, this is not to say that other combinations of elements might not result in coherent forms of realism; indeed, I suspect that other combinations may well be defensible. I do hope that nevertheless, where the motivations for the various accounts I have given of these elements resonate with readers, they will find the framework offered here provocative, and perhaps useful.

I would like to end by reaching out, briefly, from the perspective of realism, to build a bridge of sorts to other perspectives on the nature of scientific knowledge in the philosophy of science. In Chapter 1, I divided the main antirealist positions into several categories, including constructive empiricism, logical positivism and empiricism, and instrumentalism. Though in different ways, each of these views opposes realism by privileging beliefs about observables, and by withholding belief from claims about unobservables, literally construed. This is a conflict regarding what precisely one should take to be the epistemic output of the sciences: whether it includes both surface facts about observable phenomena and underlying accounts of unobservable entities and processes, or just the former.

My efforts in this book have focused on disputes concerning the unobservable in this conflict, but it would be a serious mistake to overlook the fact that even the most ardent realist must grapple with the significance of the observable. Of course, realists regard observable consequences as furnishing important tests of the truth of theories, but I intend something different here. It is also the case that in scientific practice, one is often primarily concerned with whether and to what extent theories, models, procedures, tests, etc. *work* – to enable us to cure diseases, send space shuttles to space stations, and to complete successfully the astounding variety of more humble tasks associated with laboratories and fieldwork across the globe on a daily basis. Success in practice is assessed by means of observable consequences. There is a strong current of pragmatism built into everyday scientific pursuits. The pragmatist's test of epistemic significance is utility, and utility is measured in terms of observables.

Are realist interpretations of scientific knowledge thus out of touch with what really matters to science in the real world, as opposed to the rarefied philosophical worlds of imagined science? Given the prevalence of empirically adequate idealizations and pure abstractions used in the manner of idealizations, many antirealists have assumed, I think, that realism is out of touch. It is for this reason that the notion of approximate truth, and more specifically, the unifying strand of this chapter that different sorts of truths may be contained within different sorts of scientific representations, is so important to realism. Understanding the truth content of both idealizations and pure abstractions applied to systems they do not correctly describe, the realist is able to connect motivations that many antirealists believe are independent of one another: the desire to generate observable predictions within acceptable margins of error (the driving force behind much scientific endeavour); and the desire to uncover facts regarding unobservables that underlie these predictions. Earlier I suggested that a consideration of two important features of scientific knowledge would facilitate an understanding of the concept of approximate truth. The first of these is the distinction between abstraction and idealization, and the second concerns the pragmatic dimensions of scientific practice. These points, it would seem, are intimately connected. Having considered the first, let me now move on to the second, by means of a second analogy to representation in art.

The history of twentieth-century art is in large measure a history of the avant-gardes and their forms of "abstraction". Realistic conventions of representation, in Goodman's sense, gave way to varieties of experiments that sought to realize different sorts of conventions, both in the service of representation and even, in some cases, in pursuit of *non*-representational expression. These experiments initiated the now familiar traditions of Cubism, Surrealism, Constructivism, and Abstract Expressionism. If one were to search for anything like a unifying theme in the art of the avant-gardes, however, it would be found perhaps in the increasing focus on processes of art production, as opposed to anything concerning the visual properties of the products of such processes. Many of these artists were self-consciously and primarily interested in considering the nature of artistic representation itself, paying great attention, for example, to the nature of the canvas as a two-dimensional surface, as opposed to concentrating first and foremost on the previously assumed task of realistically representing three-dimensional subjects. This is one, partial interpretation of the motivations of analytic Cubism, but it is also a recurring theme elsewhere. In the Russian tradition, for instance, Malevich's Suprematism emphasized the materiality of the process of painting – the surface of the canvas, its shape, the thickness of the paint, and so on – as opposed to traditional concerns about realistic representation. To co-opt a slogan coined by the American art critic Clement Greenberg (2003/1939, p. 539), this is 'art for art's sake'.

Perhaps this tendency towards attaching greater significance to processes involved in the creation of art as opposed to its products *per se* has its ultimate expression in the development of performance art. Consider works associated with the Fluxus movement, such as Yoko Ono's *Cut Piece*, which was performed twice, once in Tokyo (1964) and once in New York (1965). During each of these events, the artist sat on a stage while members of the audience approached, individually and in succession, to cut pieces of clothing from her body with a pair of scissors. Like all work in the performance art genre, the idea of a process takes on so much significance here that *it* now is the central focus. What matters is an event or a series of events; the notion of attaching the value of the performance to a consideration of any further output is completely lost. It is true of course that photographs of works of performance art are very important for purposes of discussion and art criticism, but these photographic traces are considered mere documents of the art form, not things that are important in their own right, and certainly not things that are the proper focus of attention when considering the nature or significance of the work.

With this suggestive idea of transition in focus from products to production in mind, let me now return to the domain of interest here, and give the pragmatic dimension of scientific practice the weight it is due. Just as focusing on processes of production led artists to a dizzying array of less than realistic representations, focusing on processes of detection, experiment, and the innumerable tasks undertaken in the course of everyday scientific work leads scientists to create ingenious abstractions and idealizations. So far as much of this work is concerned, one does not require anything like truth *simpliciter*. Abstractions and idealizations are so ubiquitous in the sciences not least because they facilitate these tasks so well, within the degrees of accuracy and precision required in particular scientific contexts. Indeed, less approximately true theories are commonly preferred to more approximately true ones. While both may generate predictions that are adequate to the tasks at hand, simpler though less approximately true theories are more easily taught, learned, and used.

In fact, the epistemic virtues of false theories often extend beyond their mere adequacy. In the early stages of this work I proposed an account of realism according to which the proper subject of realist commitment, in the first instance, is a subset of causal properties (detection properties) and certain relations between them, or concrete structures. One routinely applies pure abstractions and idealizations to phenomena whose concrete structures they do not correctly describe, but that is not to say that they cannot yield truths in such cases.[5] The classical theory of gases idealizes the nature of gas molecules and their relations to one another, but nevertheless has the (putatively) true consequences that there are molecules composing gases, and that they have properties such as mass. These are truths about particulars and properties that follow immediately from successful reference, but others arguably go further. Frictionless surfaces are ideal, but models of objects sliding down frictionless inclined planes correctly describe the motions of spherical objects as linear nonetheless, and Newtonian models of the Earth-moon system, though idealized, represent the mass of the Earth as being greater than that of the moon. Idealizations yield substantially less in the way of truth *simpliciter* than pure abstractions, but what truths they do contain may add to their pragmatic utility.

The analogy of emphasizing products versus production between art and science has, as one might expect, an echo in the intellectual traditions that scrutinize these practices. Perhaps the parallels are more suggestive than deep, but nonetheless, there is a delightful symmetry to be found in the juxtaposition of twentieth-century art criticism and post-positivist philosophy of science. One of the most constant themes to appear in critiques of logical positivism is that it is too wrapped up in normative

---

[5] Thanks to Martin Thomson-Jones and Juha Saatsi for stressing this point, and for some of the following examples.

projects based on rational reconstructions of the products of the sciences, such as theories and models, and as a consequence finds itself out of touch with the realities of scientific practice. Thus it is no surprise that the demise of positivism in the twentieth century was accompanied by the rise of the history of science as an important tool for philosophers. A great deal of post-positivist philosophy of science takes as its focus the everyday tasks of scientific practice, and correspondingly de-emphasizes the epistemic status of its products. And so the word 'truth' does not even appear in Kuhn's iconic history and philosophy of science text, *The Structure of Scientific Revolutions*, and Hacking (1983) is ultimately much more interested in intervening in the natural world than representing it.

Scientific realists should take this sort of pragmatism to heart in their understanding of approximate truth. They should think of approximate truth as something that is multiply realized by means of different representational relationships, involving true descriptions of concrete structures in some cases, and little more than successful reference in others. Some representations are purely abstract, in which case they yield all manner of true descriptions of certain classes of phenomena. Other representations are heavily idealized, in which case their truth rests in existential claims for the most part, and in the extent to which their descriptions measure up to true descriptions, in ways that are specifiable in connection with specific target systems. Of course, most cases of scientific representation are neither pure abstractions nor pure idealizations, but rather mixtures of both, in different proportions and to varying degrees. The concept of approximate truth is thus heterogeneous, to be explicated as appropriate to particular cases within the myriad contexts of representation to which it may be applied. This understanding of the truth content of scientific theories and models may be less tidy than some would have liked. Scientific knowledge is often messy, however. I hope that the informal approach to approximate truth taken here contributes towards showing how a sophisticated realism may constitute a coherent and compelling epistemic stance, regarding the amazing diversity of practices we call science.

References

Albert, D. Z. (1992). *Quantum Mechanics and Experience*. Cambridge: Harvard University Press.

Armstrong, D. M. (1983). *What Is a Law of Nature?* Cambridge: Cambridge University Press.

—— (1999). 'The Causal Theory of Properties: Properties According to Shoemaker, Ellis, and Others', *Philosophical Topics* 26: 25–37.

Armstrong, D. M., C. B. Martin, and U. T. Place (1996). *Dispositions: A Debate*, T. Crane (ed.). London: Routledge.

Aronson, J. L. (1990). 'Verisimilitude and Type Hierarchies', *Philosophical Topics* 18: 5–28.

Aronson, J. L., R. Harré, and E. C. Way (1994). *Realism Rescued: How Scientific Progress Is Possible*. London: Duckworth.

Beauchamp, T. L., and A. Rosenberg (1981). *Hume and the Problem of Causation*. Oxford: Oxford University Press.

Beckermann, A., H. Flohr, and J. Kim (eds.) (1992). *Emergence or Reduction? Essays on the Prospects of Nonreductive Physicalism*. Berlin: de Gruyter.

Bird, A. (2005). 'The Dispositionalist Conception of Laws', *Foundations of Science* 10: 353–70.

Black, R. (2000). 'Against Quidditism', *Australasian Journal of Philosophy* 78: 87–104.

Blackburn, S. (1993). 'Hume and Thick Connexions', in *Essays in Quasi-Realism*, pp. 94–107. Oxford: Oxford University Press.

Bohm, D., and B. J. Hiley (1993). *The Undivided Universe*. London: Routledge.

Boyd, R. (1981). 'Scientific Realism and Naturalistic Epistemology', in P. D. Asquith and R. N. Giere (eds.), *PSA 1980*, vol. II, pp. 613–62. East Lansing: Philosophy of Science Association.

Boyd, R. N. (1999). 'Homeostasis, Species, and Higher Taxa', in R. A. Wilson (ed.), *Species: New Interdisciplinary Essays*, pp. 141–85. Cambridge, MA: MIT Press.

Brown, H. I. (1977). *Perception, Theory and Commitment*. Chicago: University of Chicago Press.

Carnap, R. (1950). 'Empiricism, Semantics and Ontology', *Revue Intérnationale de Philosophie* 4: 20–40. Reprinted in R. Carnap 1956: *Meaning and*

Necessity: A Study in Semantic and Modal Logic. Chicago: University of Chicago Press.

Carroll, J. W. (1987). 'Ontology and the Laws of Nature', *Australasian Journal of Philosophy* 65: 261–76.

Cartwright, N. (1983). *How the Laws of Physics Lie.* Oxford: Clarendon Press.

(1989). *Nature's Capacities and Their Measurement.* Oxford: Clarendon Press.

Castellani, E. (ed.) (1998). *Interpreting Bodies: Classical and Quantum Objects in Modern Physics.* Princeton: Princeton University Press.

Chakravartty, A. (2001). 'Getting Real With Quanta', review of C. Norris, *Quantum Theory and the Flight from Realism: Philosophical Responses to Quantum Mechanics, Metascience* 10: 483–7.

Costa, M. J. (1989). 'Hume and Causal Realism', *Australasian Journal of Philosophy* 67: 172–90.

Craig, E. (1987). *The Mind of God and the Words of Man.* Oxford: Clarendon Press.

(2000). 'Hume on Causality – Projectivist *and* Realist?', in R. Read and K. A. Richman (eds.), *The New Hume Debate*, pp. 113–21. London: Routledge.

Davidson, D. (1980). *Essays on Actions and Events.* Oxford: Clarendon Press.

de Sousa, R. (1989). 'Kinds of Kinds: Individuality and Biological Species', *International Studies in the Philosophy of Science* 3: 119–35.

Demopoulos, W., and M. Friedman (1985). 'Critical Notice: Bertrand Russell's *The Analysis of Matter*: Its Historical Context and Contemporary Interest', *Philosophy of Science* 52: 621–39.

Devitt, M. (1991). *Realism and Truth.* Oxford: Blackwell.

Dowe, P. (1992). 'Wesley Salmon's Process Theory of Causality and the Conserved Quantity Theory', *Philosophy of Science* 59: 195–216.

(2000). *Physical Causation.* Cambridge: Cambridge University Press.

Dretske, F. I. (1977). 'Laws of Nature', *Philosophy of Science* 44: 248–68.

Dupré, J. (1993). *The Disorder of Things: Metaphysical Foundations of the Disunity of Science.* Cambridge, MA: Harvard University Press.

Ehring, D. (1997). *Causation and Persistence.* Oxford: Oxford University Press.

Elder, C. L. (1994). 'Laws, Natures, and Contingent Necessities', *Philosophy and Phenomenological Research* 54: 649–67.

Ellis, B. (1999). Contributions to H. Sankey (ed.), *Causation and Laws of Nature.* Dordrecht: Kluwer.

(2000). 'Causal Laws and Singular Causation', *Philosophy and Phenomenological Research* 61: 329–51.

(2001). *Scientific Essentialism.* Cambridge: Cambridge University Press.

Ellis, B., and C. Lierse (1994). 'Dispositional Essentialism', *Australasian Journal of Philosophy* 72: 27–45.

English, J. (1973). 'Underdetermination: Craig and Ramsey', *Journal of Philosophy* 70: 453–62.

Ereshefsky, M. (1992). 'Eliminative Pluralism', *Philosophy of Science* 59: 671–90.

(1998). 'Species Pluralism and Anti–Realism', *Philosophy of Science* 65: 103–20.

(2001). *The Poverty of the Linnaean Hierarchy*. Cambridge: Cambridge University Press.

Ereshefsky, M., and M. Matthen (2005). 'Taxonomy, Polymorphism, and History: An Introduction to Population Structure Theory', *Philosophy of Science* 72: 1–21.

Fales, E. (1990). *Causation and Universals*. London: Routledge.

Fine, A. (1996). *The Shaky Game: Einstein and the Quantum Theory*, 2nd edn. Chicago: University of Chicago Press.

Franklin, A. (1986). *The Neglect of Experiment*. Cambridge: Cambridge University Press.

(1990). *Experiment, Right or Wrong*. Cambridge: Cambridge University Press.

French, S. (1989). 'Identity and Individuality in Classical and Quantum Physics', *Australasian Journal of Philosophy* 67: 432–46.

(1998). 'On the Withering Away of Physical Objects', in E. Castellani (ed.), *Interpreting Bodies: Classical and Quantum Objects in Modern Physics*, pp. 93–113. Princeton: Princeton University Press.

(1999). 'Models and Mathematics in Physics: The Role of Group Theory', in J. Butterfield and C. Pagonis (eds.), *From Physics to Philosophy*, pp. 187–207. Cambridge: Cambridge University Press.

(2003). 'Scribbling on the Blank Sheet: Eddington's Structuralist Conception of Objects', *Studies In History and Philosophy of Modern Physics*, 34: 227–59.

French, S., and H. Kamminga (eds.) (1993). *Correspondence, Invariance and Heuristics*. Dordrecht: Kluwer.

French, S., and J. Ladyman (1999). 'Reinflating the Semantic Approach', *International Studies in the Philosophy of Science* 13: 103–21.

(2003). 'Remodeling Structural Realism: Quantum Mechanics and the Metaphysics of Structure', *Synthese* 136: 31–56.

French, S., and M. L. G. Redhead (1988). 'Quantum Physics and the Identity of Indiscernibles', *British Journal for the Philosophy of Science* 39: 233–46.

Ghiselin, M. T. (1974). 'A Radical Solution to the Species Problem', *Systematic Zoology* 23: 536–44.

Giere, R. N. (1988). *Explaining Science: A Cognitive Approach*. Chicago: University of Chicago Press.

Goodman, N. (1976). *Languages of Art: An Approach to a Theory of Symbols*, 2nd edn. Indianapolis: Hackett.

Gower, B. (2000). 'Cassirer, Schlick and "Structural" Realism: The Philosophy of the Exact Sciences in the Background to Early Logical Positivism', *British Journal for the History of Science* 8: 71–106.

Greenberg, C. (2003/1939). 'Avant-Garde and Kitsch', in C. Harrison and P. Wood (eds.), *Art in Theory, 1900–2000: An Anthology of Changing Ideas*. Oxford: Blackwell.

Griffiths, P. E. (1999). 'Squaring the Circle: Natural Kinds with Historical Essences', in R. A. Wilson (ed.), *Species: New Interdisciplinary Essays*, pp. 209–28. Cambridge, MA: MIT Press.

Hacking, I. (1982). 'Experimentation and Scientific Realism', *Philosophical Topics* 13: 71–87.

(1983). *Representing and Intervening*. Cambridge: Cambridge University Press.

(1991). 'A Tradition of Natural Kinds', *Philosophical Studies* 61: 109–26.

(2007). 'Natural Kinds: Rosy Dawn, Scholastic Twilight', in A. O'Hear (ed.), *The Philosophy of Science*, pp. 203–39. Cambridge: Cambridge University Press.

(forthcoming): *The Tradition of Natural Kinds*.

Hardin, C. L., and A. Rosenberg (1982). 'In Defence of Convergent Realism', *Philosophy of Science* 49: 604–15.

Horwich, P. (1987). *Asymmetries in Time: Problems in the Philosophy of Science*. Cambridge, MA: MIT Press.

Huggett, N. (1997). 'Identity, Quantum Mechanics and Common Sense', *Monist* 80: 118–30.

Hull, D. (1976). 'Are Species Really Individuals?', *Systematic Zoology* 25: 174–91.

(1978). 'A Matter of Individuality', *Philosophy of Science* 45: 335–60.

Hume, D. (1975/1777). *Enquiries Concerning Human Understanding and Concerning the Principles of Morals*, P. H. Nidditch (ed.). Oxford: Clarendon Press.

Hurewicz, W., and H. Wallman (1969/1948). *Dimension Theory*. Princeton: Princeton University Press.

Hylton, P. (1990). *Russell, Idealism, and the Emergence of Analytic Philosophy*. Oxford: Clarendon.

Jones, M. R. (2005). 'Idealization and Abstraction: A Framework', in M. R. Jones and N. Cartwright (eds.), *Idealization XII: Correcting the Model, Poznań Studies in the Philosophy of the Sciences and the Humanities* 86: 173–217.

Khalidi, K. A. (1993). 'Carving Nature at the Joints', *Philosophy of Science* 60: 100–13.

King, J. C. (1995). 'Structured Propositions and Complex Predicates', *Noûs* 29: 516–35.

Kitcher, P. (1984). 'Species', *Philosophy of Science* 51: 308–33.

(1993). *The Advancement of Science: Science Without Legend, Objectivity without Illusions*. Oxford: Oxford University Press.

Kornblith, H. (1993). *Inductive Inference and Its Natural Ground: An Essay in Naturalistic Epistemology*. Cambridge, MA: MIT Press.

Kripke, S. A. (1980). *Naming and Necessity*. Oxford: Blackwell.

Kuhn, T. S. (1970/1962). *The Structure of Scientific Revolutions*. Chicago: University of Chicago Press.

(1977/1971). 'Concepts of Cause in the Development of Physics', in *The Essential Tension: Selected Studies in Scientific Tradition and Change*, pp. 21–30. Chicago: University of Chicago Press.

Kukla, A. (1998). *Studies in Scientific Realism*. Oxford: Oxford University Press.

Ladyman, J. (1998). 'What is Structural Realism?', *Studies in History and Philosophy of Science* 29: 409–24.

Laudan, L. (1981). 'A Confutation of Convergent Realism', *Philosophy of Science* 48: 19–48. Reprinted in D. Papineau (ed.) (1996). *The Philosophy of Science*, pp. 107–38. Oxford: Oxford University Press.

Leplin, J. (1997). *A Novel Defence of Scientific Realism*. Oxford: Oxford University Press.

Lewis, D. (1973). 'Causation', *Journal of Philosophy* 70: 556–67.

(1983). 'New Work for a Theory of Universals', *Australasian Journal of Philosophy* 61: 343–77.

Lipton, P. (1993). 'Is the Best Good Enough?', *Proceedings of the Aristotelian Society* 93: 89–104.

(1999). 'All Else Being Equal', *Philosophy* 74: 155–68.

Locke, J. (1975/1689). *An Essay Concerning Human Understanding*, P. H. Nidditch (ed.). Oxford: Oxford University Press.

Mackie, J. L. (1965). 'Causes and Conditions', *American Philosophical Quarterly* 2: 245–64.

McMullin, E. (1985). 'Galilean Idealization', *Studies in History and Philosophy of Science* 16: 247–73.

Mates, B. (trans.) (1996). *The Skeptic Way: Sextus Empiricus's Outlines of Pyrrhonism*. Oxford: Oxford University Press.

Maxwell, G. (1970a). 'Theories, Perception, and Structural Realism', in R. G. Colodny (ed.), *The Nature and Function of Scientific Theories*, pp. 3–34. Pittsburgh: University of Pittsburgh Press.

(1970b). 'Structural Realism and the Meaning of Theoretical Terms', in M. Radner and S. Winokur (eds.), *Minnesota Studies in the Philosophy of Science*, vol. IV. Minneapolis: University of Minnesota Press.

Mellor, D. H. (1991/1974). 'In Defence of Dispositions', in *Matters of Metaphysics*, pp. 104–22. Cambridge: Cambridge University Press. Originally published in *Philosophical Review* 83: 157–81.

(1991/1980). 'Necessities and Universals in Natural Laws', in *Matters of Metaphysics*, pp. 136–53. Cambridge: Cambridge University Press. Originally published in D. H. Mellor (ed.), *Science, Belief and Behaviour*, pp. 105–25. Cambridge: Cambridge University Press.

(1995). *The Facts of Causation*. London: Routledge.

Mellor, D. H., and T. Crane (1991/1990). 'There is No Question of Physicalism', in D. H. Mellor: *Matters of Metaphysics*, pp. 82–103. Cambridge: Cambridge University Press. Originally published in *Mind* 99: 185–206.

Mill, J. S. (1846). *A System of Logic*. New York: Harper.

Miller, D. (1974). 'Popper's Qualitative Theory of Verisimilitude', *British Journal for the Philosophy of Science* 25: 166–77.

(1976). 'Verisimilitude Redeflated', *British Journal for the Philosophy of Science* 27: 363–80.

Morrison, M. (1990). 'Theory, Intervention and Realism', *Synthese* 82: 1–22.

Mumford, S. (1995). 'Ellis and Lierse on Dispositional Essentialism', *Australasian Journal of Philosophy* 73: 606–12.

(1998). *Dispositions*. Oxford: Clarendon Press.

Newman, M. H. A. (1928). 'Mr. Russell's "Causal Theory of Perception"', *Mind* 37: 137–48.

Niiniluoto, I. (1984). *Is Science Progressive?* Dordrecht: D. Reidel.

—— (1987). *Truthlikeness*. Dordrecht: D. Reidel.

—— (1998). 'Verisimilitude: The Third Period', *British Journal for the Philosophy of Science* 49: 1–29.

—— (1999). *Critical Scientific Realism*. Oxford: Clarendon Press.

Oddie, G. (1986a). 'The Poverty of the Popperian Program for Truthlikeness', *Philosophy of Science* 53: 163–78.

—— (1986b). *Likeness to Truth*. Dordrecht: D. Reidel.

—— (1990). 'Verisimilitude by Power Relations', *British Journal for the Philosophy of Science* 41: 129–35.

Okasha, S. (2002). 'Darwinian Metaphysics: Species and the Question of Essentialism', *Synthese* 131: 191–213.

Papineau, D. (ed.) (1996). *The Philosophy of Science*. Oxford: Oxford University Press.

Pearl, J. (2000). *Causality: Models, Reasoning, and Inference*. Cambridge: Cambridge University Press.

Poincaré, H. (1952/1905). *Science and Hypothesis*. New York: Dover.

Popper, K. R. (1972). *Conjectures and Refutations: The Growth of Knowledge*, 4th edn. London: Routledge and Kegan Paul.

Post, H. R. (1971). 'Correspondence, Invariance and Heuristics: In Praise of Conservative Induction', *Studies in History and Philosophy of Science* 2: 213–55.

Psillos, S. (1995). 'Is Structural Realism the Best of Both Worlds?', *Dialectica* 49: 15–46.

—— (1999). *Scientific Realism: How Science Tracks Truth*. London: Routledge.

—— (2001). 'Is Structural Realism Possible?', *Philosophy of Science Proceedings*, 68: S13–S24.

Putnam, H. (1975). *Mathematics, Matter and Method*. Cambridge: Cambridge University Press.

—— (1978). *Meaning and the Moral Sciences*. London: Routledge.

Quine, W. V. O. (1976). 'Whither Physical Objects?', in R. S. Cohen, P. K. Feyerabend, and M. W. Wartofsky (eds.), *Essays in Memory of Imre Lakatos*, pp. 497–504. Dordrecht: D. Reidel.

Redhead, M. L. G. (1984). 'Unification in Science', review of C. F. von Weizsäcker, *The Unity of Nature*, *British Journal for the Philosophy of Science* 35: 274–79.

—— (2001a). 'Quests of a Realist', review of S. Psillos 1999, *Scientific Realism: How Science Tracks Truth*, in *Metascience* 10: 341–7.

—— (2001b). 'The Intelligibility of the Universe', *Philosophy* 48 (Supplementary): 73–90.

Redhead, M. L. G., and P. Teller (1992). 'Particle Labels and the Theory of Indistinguishable Particles in Quantum Mechanics', *British Journal for the Philosophy of Science* 43: 201–18.

Rosenberg, A. (1984). 'Mackie and Shoemaker on Dispositions and Properties', *Midwest Studies in Philosophy* 9: 77–91.

Ruse, M. (1987). 'Biological Species: Natural Kinds, Individuals, or What?', *British Journal for the Philosophy of Science* 38: 225–42.

Russell, B. (1927). *The Analysis of Matter*. London: Kegan Paul, Trench, Trubner and Co.

—— (1948). *Human Knowledge: Its Scope and Limits*. London: George Allen and Unwin.

—— (1953/1918). 'On the Notion of Cause', in *Mysticism and Logic*, pp. 171–96. London: Penguin. Originally published in *Proceedings of the Aristotelian Society*, 1912–13.

Ryle, G. (1949). *The Concept of Mind*. Chicago: University of Chicago Press.

Saatsi, J. (2005). 'Reconsidering the Fresnel-Maxwell Theory Shift: How the Realist Can Have Her Cake and EAT It Too', *Studies in History and Philosophy of Science* 36: 509–38.

Salmon, W. C. (1984). *Scientific Explanation and the Causal Structure of the World*. Princeton: Princeton University Press.

—— (1997). 'Causality and Explanation: A Reply to Two Critics', *Philosophy of Science* 64: 461–77.

—— (1998/1994). 'Causality without Counterfactuals', in *Causality and Explanation*, pp. 248–60. Oxford: Oxford University Press. Originally published in *Philosophy of Science* 61: 297–312.

—— (1998). *Causality and Explanation*. Oxford: Oxford University Press.

Sankey, H. (1997). 'Induction and Natural Kinds', *Principia* 1: 235–54.

Schrödinger, E. (1967). *What Is Life?: The Physical Aspect of the Living Cell, with Mind and Matter and Autobiographical Sketches*. Cambridge: Cambridge University Press.

Shapere, D. (1982). 'The Concept of Observation in Science and Philosophy', *Philosophy of Science* 49: 485–525.

Shoemaker, S. (1980). 'Causality and Properties', in P. van Inwagen (ed.), *Time and Cause: Essays Presented to Richard Taylor*, pp. 109–35. Dordrecht: D. Reidel. Reprinted in D. H. Mellor and A. Oliver (eds.) 1997: *Properties*, pp. 228–54. Oxford: Oxford University Press.

—— (1998). 'Causal and Metaphysical Necessity', *Pacific Philosophical Quarterly* 79: 59–77.

Smith, P. (1998). 'Approximate Truth and Dynamical Theories', *British Journal for the Philosophy of Science* 49: 253–77.

Sober, E. (1980). 'Evolution, Population Thinking, and Essentialism', *Philosophy of Science* 47: 350–83.

Spirtes, P., C. Glymour, and R. Scheines (2001). *Causation, Prediction, and Search*, 2nd edn. Cambridge, MA: MIT Press.

Stanford, P. K. (1995). 'For Pluralism and Against Realism About Species', *Philosophy of Science* 62: 70–91.

Strawson, G. (1989). *The Secret Connexion*. Oxford: Clarendon Press.

Suárez, M. (2003). 'Scientific Representation: Against Similarity and Isomorphism', *International Studies in the Philosophy of Science* 17: 225–44.

Suppe, F. (1989). *The Semantic Conception of Theories and Scientific Realism.* Chicago: University of Illinois Press.

Swinburne, R. G. (1980). 'Properties, Causation, and Projectibility: Reply to Shoemaker', in L. J. Cohen and M. Hesse (eds.), *Applications of Inductive Logic*, pp. 313–20. Oxford: Clarendon Press.

Swoyer, C. (1982). 'The Nature of Natural Laws', *Australasian Journal of Philosophy* 60: 203–23.

Teller, P. (1995). *An Interpretative Introduction to Quantum Field Theory.* Princeton: Princeton University Press.

(2001a). 'The Ins and Outs of Counterfactual Switching', *Noûs* 35: 365–93.

(2001b). 'Twilight of the Perfect Model Model', *Erkenntnis* 55: 393–415.

(2005). 'Discussion – What is a Stance?', *Philosophical Studies* 121: 159–70.

Tichý, P. (1974). 'On Popper's Definitions of Verisimilitude', *British Journal for the Philosophy of Science* 25: 155–60.

(1976). 'Verisimilitude Redefined', *British Journal for the Philosophy of Science* 27: 25–42.

(1978). 'Verisimilitude Revisited', *Synthese* 38: 175–96.

Tooley, M. (1977). 'The Nature of Laws', *Canadian Journal of Philosophy* 7: 667–98.

(1987). *Causation: A Realist Approach.* Oxford: Clarendon Press.

van Fraassen, B. C. (1980). *The Scientific Image.* Oxford: Oxford University Press.

(1985). 'Empiricism in the Philosophy of Science', in P. M. Churchland and C. A. Hooker (eds.), *Images of Science: Essays on Realism and Empiricism, with a Reply from Bas C. van Fraassen*, pp. 245–308. Chicago: University of Chicago Press.

(1989). *Laws and Symmetry.* Oxford: Clarendon Press.

(1991). *Quantum Mechanics: An Empiricist View.* Oxford: Clarendon Press.

(2000). 'The False Hopes of Traditional Epistemology', *Philosophy and Phenomenological Research* 60: 253–80.

(2002). *The Empirical Stance.* New Haven: Yale University Press.

(2005). 'Replies to Discussion on *The Empirical Stance*', *Philosophical Studies* 121: 171–92.

Vendler, Z. (1967). 'Causal Relations', *Journal of Philosophy* 64: 704–13.

Wilkerson, T. E. (1995). *Natural Kinds.* Aldershot: Avebury.

Woodward, J. (1992). 'Realism About Laws', *Erkenntnis* 36: 181–218.

(2003). *Making Things Happen: A Theory of Causal Explanation.* Oxford: Oxford University Press.

Worrall, J. (1989). 'Structural Realism: The Best of Both Worlds?', *Dialectica* 43: 99–124. Reprinted in D. Papineau (ed.) 1996: *The Philosophy of Science*, pp. 139–65. Oxford: Oxford University Press.

(1994). 'How to Remain (Reasonably) Optimistic: Scientific Realism and the "Luminiferous Ether"', in D. Hull, M. Forbes, and M. Burian (eds.), *PSA 1994*, vol. I, pp. 334–42. East Lansing: Philosophy of Science Association.

Worrall, J., and E. G. Zahar (2001). 'Ramseyfication and Structural Realism', in E. G. Zahar, *Poincaré's Philosophy: From Conventionalism to Phenomenology*, Appendix IV, pp. 236–51. Chicago: Open Court.

Zahar, E. G. (1996). 'Poincaré's Structural Realism and His Logic of Discovery', in J.-L. Greffe, G. Heinzmann, and K. Lorenz (eds.), *Henri Poincaré: Science and Philosophy*, pp. 45–68. Berlin: Akademie Verlag and A. Blanchard.

Index